The Simple Guide to Home Electronics 2017

The Simple Guide to Home Electronics 2017

Lightbulbs, TVs, Internet & more
in plain language

Mark Schutte

Suwanee Press

2017

First Printing: May 2017

ISBN 978-1-947333-00-0

Suwanee Press
1115 River Laurel Dr.
Suwanee, Georgia, 30024

www.SuwaneePress.co

This book is dedicated to my mother

Dorothy Griggs Schutte

Who nudged me towards a career in electrical
engineering which has been fun and rewarding

&

My wife

Lisa P. Schutte

For her encouragement and insightful comments

Table of Contents

Preface

We were having dinner with our neighbors. The company was grand, and the food and beverages delicious. Hank was telling us about his new TV setup with Direct TV. But his wife had "hit a button" and now neither one could get TV channels on the upstairs TV. Hank and Kim are college graduates. They are world travelers and very successful in business. I'm describing very intelligent people. And yet, Hank was dreading calling Direct TV for help because he would feel foolish. After all, it was working when the installer left three days before. What could be wrong? I offered to come over and see if I could sort it out in exchange for a beer and a tour of their new house. We needed to visit anyway. If you understand today's gadgets, the problem was simple. It took me 30 seconds to fix it and 30 minutes to write down how to get out of this situation the next time the wrong button was pressed. Today's gadgets can be intimidating to many of us. And very often confusing to the rest of us.

Most of us used to walk into the electronics store and buy a new gadget with confidence. We did not have to find a teenager to help us hook it up. Now we often go into the store with the same dread as buying a car. What do all those words mean anyway? Does the 16-year-old salesperson really know what they are talking about. Sadly, the answer is often, "No". And we feel dumb. For my friends, relatives and neighbors and the rest of the world full of good folks just like us, I've attempted to write a simple guide to help arm you with information.

My lovely wife did not know what kind of light bulb to buy. Incandescent, CFL, LED or the more exotic halogen? How hard can it be?

Humans learn new things on a somewhat linear scale. We can adapt to change if it is handed to us slowly enough. The problem is that technology is changing exponentially and with leaps and bounds. Moore's Law has the number of transistors on a chip doubling every 18 months. In the early years, this was change we could adapt to. Now going from 10 million transistors to 20 million transistors enables real science fiction leaps of change. Today, the government is trying to figure out how to regulate Uber, the car service. At the current course and speed, by the time they figure this out, Uber will be made obsolete by self-driving cars. Not only people but society and governments are struggling to keep up.

But, have no fear! With a little information, modern gadgets are not all that complicated. This book seeks to provide that information for all the folks that don't have us over for dinner on a regular basis. I am an electrical engineer. More than that, I've been in the consumer electronics industry for more than a few decades. My struggle is to describe it in a simple way that you can understand. To do this I will hit the high points. I will use non-technical terms. The descriptions will lack the normal rigor that might cause you to go comatose. I will use normal English, tell stories, and get the important points across without the reader needing a technical education to understand.

I will talk about real products. I have not gotten paid by any of these companies. If I like it, I say so and say why. I haven't used or tested everything. Something else may be better. But when I recommend something there is a reason. Some products I've studied and reviewed, but don't have first-hand experience with. I can't play with everything since I'm using my own money. But more importantly, too much technology in our home causes more than a bit of stress with half of the family and I want to stay married to the same woman who has put up with me thus far.

> One of my rules for success is not to be afraid of appearing stupid. If someone throws a "TLA" *[Three Letter Abbreviation]* at you, stop them and ask, "What does that mean?" It is often a technical term not understood outside a small group of professionals.
>
> Question the advice from someone that cannot explain their TLAs.

This is not a "Dummies" book. Everyone struggles with some parts of the material covered. There is too much new stuff to keep up. I'd say it was targeted at the over 30 crowd, but with new TV standards like UHD, HDR and ATSC 3.0, *[These are explained in the TV chapter.]* frankly few people outside the industry really know what is going on and where the pitfalls are. So be assured, you are not alone. We all struggle.

I'll start each chapter with the answer, to what my old boss Ken used to call the "So What" question. If you don't read anything else, this can get you to a reasonable answer for most situations. However, "your mileage may vary". Meaning, your situation may be different and if you read more, you might find a better answer for your exact situation. The rest of the chapter will tell you why the answer works and what assumptions I made in choosing it. And you might learn something along the way. But have fun with it.

There is a web site associated with this book. http://www.suwaneepress.com/simplehome. There is a link to an email address for questions. questions@suwaneepress.com If you think of a question that I don't cover, I'll try to answer it. If many people ask the same question, I'll post the answer on the web site and update the next edition of the book. The book has a date. Things are changing even as I write it so to stay current, I'll likely have to revise it periodically.

Don't read this book cover to cover. I mean, you can if you want. My dad once set out to read the entire 26 volume Encyclopedia Britannica. I suspect he was just hitting the high points of some articles, but I don't think he made it past "M".

Look at the Table of Contents. Read the chapter you need to do the task at hand. Many of the topics are inter-related. There is some thought to the order of the book. But where you need to know about Internet to hook up your TV, there will be a pointer or link to that section. So, you may jump around a bit. Frankly the Internet touches most everything today and it is covered early in the book.

The simple illustrations are mine. I drew them. Engineers have a hard time talking without using their hands or a white board. The photos and cool drawings are from strangers who generously listed them for public use in the creative commons. Many thanks to those folks for generously helping bring clarity to the subject.

Introduction

There was a time when there were only 3 TV networks. Music came from the radio, was performed live or was purchased on vinyl records. Photography was on film with a camera. Your film was most likely Kodak. Phones were tied to wires in the home and if you wanted to make a call while driving you found a payphone by the side of the road. Long distance calls were expensive. Mail got delivered to your home on paper. You shopped in stores and used paper catalogs to find what was available. Movies were at the theater and viewed in the company of a hundred or so strangers. Information was found in the library or in the volumes of an encyclopedia.

Today there are 500+ channels (and fewer programs you want to watch). TV is also moving from broadcast to on-demand streaming (aka Netflix, Hulu, SlingTV, Direct TV Now and Amazon). Music is on CD but is moving to streaming services like Spotify, Pandora, Apple and Amazon. You likely take photos with your cell phone. You cannot find a phone booth. Kodak no longer exists as the premier photography company. You likely communicate with friends and family with email or text messages and only bills come on paper in *snail mail*. Even bills want to go *digital*. Amazon is replacing the department store and packages show up at your door in a few days. Amazon recently filed a patent for an orbiting warehouse and delivery by drone in a few hours. Movies can easily be seen on your big screen TV. The 'web' has replaced catalogs and you find information in Wikipedia. And then there are social media such as Facebook, Instagram and Twitter.

The world has changed. Previous generations saw the change from horses to automobiles. Manned flight was achieved; telephones and electricity became commonplace in the home. There is a considerable body of research showing that as invention builds on top of invention, the rate of change is exponential. We as humans adapt to change in a more limited linear fashion. Regardless, the change over the last 40 years has come with its own new vocabulary and can seem strange and foreign to those of us of a certain age. We feel dazed and confused as we talk to our peers, co-workers, children and grandchildren. Our banks and even social security want us to interact with them differently. They assume we have a smart phone.

It is not only people who struggle with technology change. It can affect organizations like Social Security. They recently moved to require that people communicate with them via a smart phone. My mother does not have a smart phone. This is apparently a common problem and Social Security has wisely backed down.

If we want to read a story to a grandchild living in a distant state, there is something called Skype or Facetime or Google+. Since we are not ready to join the Amish, or leave this world, we need to sort some of this stuff out. In fact, simply buying a replacement light bulb presents us with a dizzying array of choices.

Join me in the adventure of trying to make home electronics simple.

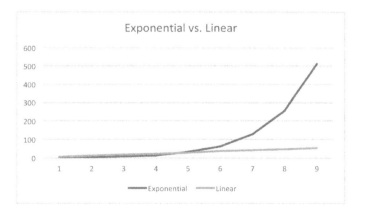

Chapter 1 Lights

The answer is: LED

If you are buying a bulb to be used for any length of time, LED is likely the most cost effective and reliable.

The purpose of light bulbs is to produce light. That seems self-evident but it is important. What we want out of our light bulbs is illumination. A decade or so ago, you specified a light bulb by the power it consumed. For example: a 7 W (watt) bulb was a night light. A 40 W was dim. A 60 W was average. A 100 W was bright for reading and so on. Power is measured in watts. But the power used does not tell us anything about light. It tells us more about our electric bill. We care about two things with light; the amount or brightness, and most of us care about the color.

The color was more obvious when you took photographs on film with no flash. Indoor lighting made everything a bit red. Fluorescent lighting made things look green. Outdoors in daylight, the pictures looked good. But the brain is an amazing thing. We don't see those colors with our eyes normally. Film did not adjust itself for the color of light but your brain did. Indoor lighting did not appear red or green to you. Your brain adjusted, mostly. Sensitive people that stopped and thought about it would say this light is warm (reddish). Or they could say this light is cold (blueish) or you look ill in this light (green). But thanks to our amazing brain we can get along without problems. So, of the two qualities of light, color is minor. And by the way, modern digital cameras and even the camera on your phone do the same thing as the brain, they correct for the color of light. Minor doesn't mean unimportant. But you will typically be reduced to two choices, *Soft & Warm* or *Daylight*.

The real key to light is the amount, how bright it is. We relate the brightness of light to the power. More power equals more light. And this is true for any one type of light bulb. But not true when we compare different kinds of light bulbs. To really talk about the brightness of light, we need to introduce a new word: *Lumens*.

It is useful to talk about lumens rather than watts when we want to talk about brightness. A 60 W traditional or *incandescent* light bulb produces about 800 lumens. When you look at bulbs in the store you might see "lumens" in fine print but you are more likely to see "replaces a 60 W bulb, uses 17 W". Confusing yes? Because for a long time, consumers only had one type of light bulb and it was easier to talk watts than lumens. So now the marketing people are trying to tell you how much light a bulb puts out by comparing it to the older technology. I guess they think you can't learn a new word, "lumens." The word for this chapter is "lumens".

The brightness of light is measured in "lumens"

800 lumens is the light produced by a traditional (incandescent) 60 W light bulb. See, you feel smarter already. Now you have the vocabulary to compare and be smart about different bulbs.

There are three basic types of light bulbs in common use in 2017.

Cost based on 800 lumen light bulb			
	Traditional Incandescent	Fluorescent & Compact Fluorescent	LED
Price	$0.41	$0.99	$5.56
Watts	60	35	9.5
Lifespan (hours)	1,000	10,000	25,000
Total Cost over 20 yrs @ 13 cents/kWh	$360	$164	$86
Dimmable?	Yes	Not really	Yes
Hazardous waste	No	Yes (mercury)	No

Table 1: Comparison of light bulbs

Before I tell you about each type of bulb, take a quick peek at Table 1. If initial purchase price is the thing driving the decision, you will pick the traditional Incandescent bulb. If you have a longer investment time in mind, you will likely choose LED. In fact, let's build Table 2 for fun.

What is important	"and the winner is..."
Initial purchase price	Incandescent
Total Cost over 20 years	LED
Long Life	LED
'Green' factor (save the planet)	LED
Dimmable to get you in the mood...	Incandescent and LED

Table 2: Factors driving choice

We see that Fluorescent did not win a single factor. It used to win on total cost and lifespan for decades and then LED came along and beat it in all categories. My guess is that no new light fixtures will be built fluorescent. Bulbs will be sold to fit old fixtures for a few decades. But you should question someone wanting to install a new fluorescent fixture.

If you just wanted to know what bulb to purchase, our job is done. The government is trying to phase out incandescent bulbs because they use massive amounts of power. So, buy LED and be done with it. I personally took all the incandescent bulbs in my high ceilings and replaced them with LEDs. I figure the next time I need a bulb replaced (~15 to 20 years) I'll have a grandchild come do it for me. Read on if you must know more. But our work is really done. There is more about new features like changing color and wireless control I'll discuss later.

Incandescent

Figure 1 Edison Light Bulb
Wikipedia.org

. . .

You have all heard or read about Thomas Edison inventing the light bulb in 1879. You need to throw in some adjectives there to be correct. "practical", "commercially viable" come to mind. Inventors had been working with creating the light bulb since at least 1802. There were 22 people ahead of Edison with significant work. But his is the first one we read about as children. It had a lifespan of 13.5 hours. Lightbulbs got better. Today we get 1,000 hours. But it seems like only yesterday we changed that bulb. Today's incandescent bulb is pretty much the same as Edison's version. A bit better made. Much cheaper. But for all the history, it has significant faults.

The average 60 W incandescent bulb only turns about 2.2% of the energy (watts) into visible light. The other 97.8% is mostly turned into heat. This is great if you are trying to heat an "Easy Bake Oven". But this is bad if you want light to read by. Remember, this was state of the art for 135 years. It's time we moved on.

> My little sister had an Easy Bake oven when she was younger. The heat for the oven was produced by an incandescent light bulb. This simply goes to show that incandescent bulbs generate significant heat.

Today's incandescent bulbs have a life span of about 1,000 hours of operation. While the initial purchase cost is cheap, we called it $0.41, you would need 25 of them at $10.25 total, to equal the life of an LED bulb. Not to mention the hassle. Especially if the fixtures are hard to reach like a high ceiling.

To say something nice about the aging incandescent bulb, they are dimmable. You can create mood lighting and they have a lovely warm character. Too warm if you shoot film but that is another chapter. They also do not put hazardous chemicals in landfills when you dispose of them. You will dispose of them often.

How does it work? In simple terms, you run electricity through a thin wire in a vacuum, and it gives off radiation. Mostly heat, but a bit of visible light.

Fluorescent and Compact Fluorescent (CFL)

In the march of lighting history, fluorescent is next. Again, lots of people played with it but a company in Germany filed a patent in 1927. In the US, GE filed patents issued in 1931. It works by running electricity through low pressure mercury-vapor gas in a tube. This gives off ultraviolet (UV) light. Remember *black-lights*? But unless you are trying to bring back the psychedelic mood of your lost youth or your parents youth, ultraviolet light is not very useful. The inside of a normal fluorescent bulb is coated with a chemical and when you hit the chemical with UV light it gives off light in the range useful for lighting. The actual color depends on the chemicals that coat the inside of the tube. For film photography, they all look a little green.

Our 800-lumen fluorescent bulb takes about 35 watts compared to the 60 watts of Incandescent. So, about two times better. However, fluorescent bulbs require a regulated supply of electricity and the fixtures are more expensive.

Figure 2 CFL bulb
Wikipedia.org

Did I mention they are not easily dimmable?

What are compact fluorescent bulbs (CFL) or compact fluorescent lighting? The traditional fluorescent bulb was a long thin tube. And fluorescent lights use about half the energy of incandescent so if you could replace incandescent bulbs with fluorescent bulbs you could cut your power bill. The problem is in fitting the long thin tube that requires a regulated power supply in a plain old cheap incandescent fixture. But someone figured out that you could twist the long thin bulb into various shapes and make it fit into the same space as a bulb. Then they put the regulated power supply in the base of the bulb and Voila! But then people hated the quality of the light, the bulbs did not support three-way brightness and they were not dimmable. So, their popularity is limited. And the mercury vapor is considered a hazardous material. It should be recycled and not thrown into landfills.

LED Lighting

LED Lighting is a new thing in the last five or so years. They have no real history to speak of. And in the last couple of years the price has dropped as volumes have risen so they are now practical.

LED stands for light emitting diode. Basically, it is a solid-state device (think computer chips) that produces light when you run electricity through it. They come in red, blue and green and possibly a few other colors but not white. We all remember our color wheels from grade school; you combine colors to produce new colors. To get the shade of white light that we want for our homes we combine various numbers of the red, green and blue LEDs.

Figure 3 LED bulb
Ledlightsforhome.org

Which is good because one LED does not put off our 800 lumens. And LEDs are very directional. They emit light in one direction. But they are very cheap. So, no worries! We put in enough LEDs to get the right color and we point them in different directions and we have enough of them to generate our 800 lumens.

When I say solid-state, there are no moving parts. LEDs are very tough. They will last for a long time in a flashlight, for example. The typical life time is 25,000 hours.

The big thing with the old incandescent lights is that they come in all shapes and sizes. After 135 years, there are many different shapes available. Given the newness of LEDs, some of the shapes are not yet available. My personal experience is that manufacturers started with the most common shapes and brightness (lumens) to get volume up and price down. Recently I see more and more different shapes and sizes available. Good sources are Lowes and Home Depot. Amazon and the Internet likely have the biggest variety. But knowing what to call it when searching can be a problem. It's a bit easier to walk into a store and say, "I want an LED version of this..." and hand the clerk your old bulb.

One more thing, LEDs come in basic colors that we combine to get white. They also make LED bulbs that let you control the amount of red, green and blue to create different color lighting. These can be controlled through radio (Wi-Fi, Bluetooth, Zigbee, Z-wave and more) usually using an application (app) / program on your smart phone. This is sometimes called the Internet of Things (IoT) and we will talk about it later in the book.

Rather than replacing an older bulb with a new LED bulb, it can be easier to replace the old fixture with a new LED based fixtures. You should still pay attention to the color such as soft-warm and cool daylight.

Halogen

No, I did not forget halogen. I choose to ignore it. It sits in the middle of the other technologies and is not present in any large volume. In the future, it will be replaced by LED. If you need a halogen bulb, go buy a halogen bulb. But for anything new, go LED.

I could write about the technology but I don't want to put you any further to sleep.

The magic is in the base

I said earlier that fluorescent lights needed a constant current power supply also called a ballast. Large fluorescent fixtures have this ballast built in to the fixture. As the gas in the bulb heats up, without the ballast, it will allow more and more current to flow through it and basically burn itself up. Which is why the fixtures have a ballast to prevent this. CFL or Compact Fluorescents are designed to work in normal incandescent sockets. Where is the ballast? Simple, they hide it in the base of the bulb. Because space is limited and because you throw it away with every burned-out bulb, the ballast is constructed inexpensively. Most limit the power always, which means the bulb is very dim when you first turn it on. It only reaches full brightness later when it warms up. This is a problem in many applications like a bathroom mirror.

LED and Radio Frequency (RF) noise

LEDs do not work on household power. Household power is 120 volts Alternating Current (AC). LEDs are basically computer chips which use a Direct Current (DC). For an LED light to work it needs a power supply to convert the 120 volts AC to DC at the proper voltage. Since these bulbs are designed to work in simple incandescent fixtures the power supply is hidden in the base of the bulb. To make this power supply small and efficient the designs use a higher frequency than the usual household power. This higher frequency can have interesting side effects if the bulb is cheap and poorly designed.

I recently converted my outdoor landscape lighting to LED. The floodlights used to highlight the architectural details of my home were basically old style auto headlights. They were incandescent and got very hot. The new LED fixtures were smaller, used much less power and gave off more light. There is a power supply in each LED light unit to convert the power to DC. Apparently, the frequency of this power supply is at the same frequency of the radios used to open garage doors. The 12 LED fixtures are putting enough radio frequency (RF) noise in to the power lines to block my garage door openers in the evening when the lights are on. Judging by a few internet searches, this is not limited to a single brand of Chinese produced LED landscape light but is a problem across inexpensive LED lights in general.

I should note, I have this problem only with my landscape lights. I do not have a problem with the LED lights in my home.

Color lights may be truly be important for outdoor Street lights

Earlier I said that the color of light was minor and for every rule there of course is an exception. Given the significant savings in energy and improved longevity there is a big advantage in switching street lighting to LED. This can save cities millions of dollars in power cost and in labor to maintain these fixtures. But, recent research has shown that it is not that simple.

The early outdoor LED street lights produced a white light that was bluer in tint. While the bluer white light was driven by economics it was originally thought that blue light would help you see better at night. There is no research to prove this, but it sounded good. For home use, most users prefer a warmer or redder light and so LED manufacturers learned to produce an LED light that had more reds. But many LED street lights were still very blue.

Daylight is more blue and recent studies have shown that this light color is very key to establishing the day/night rhythm in our lives and in the lives of many animals. A blue street light can affect the day/night cycle of people and animals.

Also, traditional lighting creates light pollution near cities that makes it difficult to see the stars. Blue LED lights are significantly worse in this regard.

Manufacturers have responded by creating outdoor lighting that is much warmer and closer to the traditional street lighting that we are familiar with.

The take away is that if your community is planning to switch to LED street lighting (and you should), ensure that the newer LED lights with warmer color are selected. The technical term is "4,000 K".

The sodium vapor lights they replaced were typically 2,700 K. "K" stands for Kelvin and if you are a physicist you likely knew that. Blue is around 6,000 K for reference. I only tell you this so that you can ask for the right thing. You can forget it almost immediately.

More on Lumens

We said that you don't measure the amount of light provided by a bulb in watts. The real measurement is in lumens. To help bridge the transition between familiarity with incandescent bulbs and new lighting technologies the following table is provided.

Power of Incandescent bulb	Lumens
40 W	450 lumens
60 W	800 lumens
75 W	1,100 lumens
100 W	1,600 lumens

Chapter 2 Internet

**There is no single answer to the
Internet so read on...**

I'm covering the Internet early because so much of what follows relies on it in some form or fashion. But I really don't know where to start. So, I think I'll create a list of questions and if you already know the answer then skip that part.

- What is the Internet?
- How do I get Internet?
- How much speed is enough?
- What is the *Web*?
- What is the *Cloud*?
- What equipment do I need?
- What is Wi-Fi?
- Tell me about this new thing called Mesh.
- I'm worried about security!

What is the Internet?

People talk to each other using the phone network. Computers talk to each other using a data network. The data network in your home is called a Local Area Network (LAN). When you connect a bunch of LANs together it can be called a Wide Area Network (WAN). And all those WANS connected to each other is the Internet. So basically, the Internet is a bunch of computers talking to each other.

It started as a Defense Department Advanced Research Project (DARPA) program to allow the military's computers to talk to each other reliably. There was no central control and a message could take many paths. This is a good thing if someone is shooting missiles at you. It is still a good thing if you have normal modern day occurrences. If you are sending files to your aunt in New York and the link through Baltimore goes down, the Internet will simply send traffic through another path like Pittsburgh. When the undersea link to Asia was broken in an earthquake a few years back, the traffic was routed through Europe. Universities were let in on the Internet project and it *escaped* into industry and eventually into your home. It has changed everything in this book.

The Internet relies on a common language called Internet Protocol (IP). You can think of IP like sending a letter in the mail. The *envelope* in IP is called a packet. It has a return address and a destination address. And the packet contains the message. But computers are rather dumb beasts, which you suspected all along. The envelopes can hold no more than about 1500 characters. A long message simply uses multiple envelopes (packets). Software on the other end will put the message back together. And if an envelope is missing, the software will ask that it be sent again. These envelopes can take different routes to get to the intended recipient. This makes the Internet very robust and tolerant of failures.

How do I get Internet?

I promised the good stuff up front. In 90% of the cases your best option for high speed Internet is your local cable or phone company. But read the rest to find out why.

You will likely never know the names of the companies that run the cable (fiber optics) across oceans and around the globe. You only need to know who provides service for the *last mile*. For your home, your choice is limited by where you live. Sorry, real competition in North America has been slim to none. But it is getting better in large cities. For most of us your options are A) the cable company like Comcast, Charter, Cox or others. And B) the local phone company. Some cities like Chattanooga, TN have their own networks. And Google is laying Google Fiber in more and more cities. Some cities start to lay fiber and then find out how hard it is and they give it over to Google to run.

But we are getting ahead of ourselves. We haven't said what fiber is. Let's approach this with a historical waltz through time. Back when I was a child, your phone was connected via two wires. The cable running to your house likely had four, six or even eight wires in it. But this was in case you wanted a second, third or fourth phone line. *[This is important later]* Your Television came over the air and you had an antenna. Eventually cable TV came along and you had two cables running to your house. The phone cable and the TV cable.

Your first computer connection to the Internet used the same wires your phone used to allow you to talk. It used a device called a modem. The first were rather slow but they eventually got faster up to 56,000 bits / second. (56K). If you used a computer back then, you remember the bumps, buzzes and assorted noises they made. You were likely also connected via CompuServe or AOL.

> 'K' is shorthand for a thousand. Sometimes it is a people thousand like 1,000 and sometimes it is a computer thousand, 1,024. It related to ones and zeros, binary. Don't worry about it. Consider it round off error

More recently, you mostly have two choices, DSL from the phone company or cable modem from the cable TV company.

The two wires from the phone company are ok for voice. But it is limited for data. After working magic to squeeze more and more data into the two lines, the phone industry started using the extra pairs of wires in the cable. This is called ADSL or VDSL. *[DSL is Digital Subscriber Line.]*

This is where we say, "your mileage may vary."
Your actual speed is dependent on how close you
are to the phone company equipment. If you are
close enough and the wind is favorable, you can
get 12 to 18 Mbps. (bps is bits per second. A bit
is a one or zero.) When you get a call from the
phone company offering *high speed* data, this is
the speed they are referring to. In 2016, the FCC
changed the definition of *broadband* to be 25
Mbps or greater. The phone company cannot
offer *broadband* over the traditional
infrastructure (twisted pairs of wires) for 90% of
their customers. The phone companies recognize
this is a problem and changes are happening and
we will discuss this shortly.

The cable TV company was trying to imitate
broadcasting over the air. They used a different
cable to do this, called coax cable. It has a center
wire surrounded by an insulator (does not
conduct electricity) and a wire mesh or foil
wrapper and more insulation on the outside.
Coax can inherently carry more data. Originally it
was used for analog TV. It was divided into
channels sized for analog TV channels. You can
have one hundred or so analog TV channels on a
coax cable. When cable went digital, each TV

"K" is 1,000, "M" is 1,000,000, "G" is
1,000,000,000.] We call "k" kilo, "M" is called
mega, "G" is called giga

channel could carry 38 Mbps of data. And when

you combine several channels you get up to 1 Gbps or more.

When I started writing this book, I had 100 Mbps in my home using a cable modem. That was 100 Mbps to the home and 5 Mbps back to the Internet using a cable modem from my cable TV company.

Fiber is short for Fiber Optics. It uses laser light transmitted over a plastic or glass fiber about the thickness of a hair. It carries more data ("greater bandwidth") than coax cable. It is also more expensive to transmit and receive. The actual cable is comparatively inexpensive. The real cost is digging the ditch to install it. When you install fiber, you put in many fibers at once. This makes growth easy.

Google and many companies are using only fiber. They advertise 1 Gbps. You should understand that that Gbps is *shared* between 16 to 32 uses. And this is not a problem. If you run a speed test you will measure 1 Gbps if you have a fast-enough computer.

Because the traditional twisted pair infrastructure of the phone companies is so limited, they are moving quickly to fiber. Verizon is perhaps the most aggressive. Many of their locations are located near the ocean and the salt air is rusting out the copper wires. AT&T is starting to catch up. AT&T also agreed to speed up deployments

of fiber in exchange for permission from the
government to buy Direct TV.

*A personal note: AT&T dug up my neighborhood
yards and installed fiber (GPON, Gigabit Passive
Optical Networking). They offered me 1 Gbps in
both directions at a lower price than my 100
Mbps cable service.*

Needless to say, I switched.

Cable companies are also joining the competitive
fiber race. In more and more locations, they are
offering fiber.

The cable companies may not need to switch to
fiber to compete. There is a new cable modem
technology called DOCSIS 3.1 that can offer
speeds up to 10 Gbps or higher. Cable modems
use the coax cable already in the ground or on
the poles. And there is a newer twist to DOCSIS
3.1 called Full Duplex that adds the ability to go
fast in both directions but it is a few years out.
That coax cable installed in the 70's can give as
much or more speed as a nice new (expensive)
shiny fiber optic cable. As I said, competition is
slowly improving.

What is the Web?

The World Wide Web (WWW) or "Web" for short,
allowed normal people to join in the fun of using
computers to get information from another
computer anywhere in the world. The web is

software (a computer program). Back when computers first started talking, they used strange and unfriendly *languages* such as File Transfer Protocol (FTP) to exchange information. Tim Berners-Lee, an English Scientist working at CERN in Switzerland in 1989, was looking for an easy way to share research papers with other scientists. He created an early set of standards for a computer on one end, the server, to host a *page* and for a computer on the other end to view the page using a browser. The pages are cool because they have *links* that take you to other pages. The *web server* is one computer program. And the *web browser* is the program that runs on your computer. You call it by names such as Internet Explorer or Edge or Safari or Chrome or Firefox.

Web pages have Universal Resource Locators (URL). Ok, this is Greek. A URL is simply an address, a location. Just like an address on an envelope that you mail. Web pages generally start with some form of "www." You can omit the "www" and your browser will add it for you. Then there will be a company name like "BBC" all this is followed by a *domain* like ".com" or ".gov" and even ".org." Domains are simply categories of sites.

Specific sections of a site, like rooms in a building, get additional address information like "/news". Say you wanted to see the BBC news page, the whole address will be something like www.bbc.com/news .

Truth be told, popular web sites have multiple addresses consisting of strings of numbers to spread traffic out and give you faster service. Your local DNS server (Domain Name Server) converts the human understandable www.cnn.com and translates it to a string of numbers representing a server nearby so that you get fast response. Unless you are a networking professional needing to debug your DNS Server, you really don't need to know this. Call it a fun factoid.

Defining Web terms

Web Servers

A web server is a program running on a computer in the cloud. It serves up web pages. Which is just information written in HTML. Or Hyper Text Markup language. HTML is like the codes that Microsoft Word uses to change the font or other format codes.

Web Browsers

A web browser is a program that takes information formatted in HTML and presents it on your computer screen.

Web Domains

Think of a web domain as the country code in an address. If the Web is the whole word, A web domain is a country.

How much speed is enough?

Video is the driver for how much speed you need. Your video will come to you over the Internet soon if it does not already. A full HD movie is 18 Mbps (MPEG2, more on this in the TV section.) 4k TV requires more but it uses better compression so the growth is not proportional. You need perhaps 25+ Mbps for 4k TV. This is one TV program. If you have several people in the house watching different shows or are recording one and watching another, you need more.

I mentioned video compression. You don't need to know the details but industry has figured out how to squeeze the size of a video show down to smaller numbers. As technology advances putting more and more transistors on a chip, the chips use more complicated methods to squeeze the data smaller and smaller.

Downstream, the bits coming at you from somewhere on the Internet, is not the only consideration. *Upstream*, is the data you send back to the Internet. email, file backup, cloud syncing sends data back to the Internet. And they don't really tell you this, but pretty much all downstream data needs to be acknowledged. This is how the Internet knows if part of the data is missing and needs to be resent. Even if you are not sending anything back, you are sending data back.

I was having trouble at my vacation cabin. I had 20Mbps down/1Mbps up service. I was not watching video over the Internet at the time but even email was slow. I do this for a living and I called the local cable TV support engineer. He logged into my cable modem and said "you are maxing out your reverse traffic. Do you have any Apple devices?" The short answer is "many Apple devices." Apparently, the Apple cloud syncing uses lots of bandwidth back to the cloud. Watching TV require the box to talk to the video server and cloud syncing was blocking that conversation. We increased our package to increase our reverse bandwidth and things were fine. The reverse traffic drove the decision on which package we picked.

I think most people with two adults in the family would be OK with a 30 Mbps down and 2 Mbps upstream data plan from their cable company. If you have fiber you are likely faster than that. If you have DSL from the phone company, those speeds are not available. If you have teenagers or young adults in the house, you should think about getting a faster package like 50 Mbps downstream and 3 or more Mbps upstream. Your cable company may have different offerings and they are always getting faster. Round up from my recommendations.

What is the cloud?

Some applications use the *Cloud*. You have *cloud* storage, cloud backup, *cloud* syncing, "iCloud" and more. The cloud is something fuzzy so the name is appropriate. An engineer in my office has a handmade sign on his cube that says, "The

> *"The cloud is nothing*
> *more than somebody*
> *else's computer"*

cloud is nothing more than somebody else's computer". This is pretty close to the truth.

Back in my early childhood, computers in the movies were in big rooms with spinning tapes. There were lots of guys in white coats and clipboards. *(I don't know why they had clipboards. The ones I knew had boxes of punch cards. If you don't know what a punch card is, you are fortunate.)* Then for a time, the rooms shrank as computers got smaller and smaller. And today, the rooms are back. Stuffed with thousands of small computers all connected and talking over the Internet. The tape drives are missing and the engineers wear blue jeans and T-Shirts. These rooms with thousands of computers are the cloud. You can run a program in the cloud. You can store your files, songs and pictures in the cloud. Which machine is it on? That is a hard question to answer. It could be any one of them. The answer is fuzzy. Hence, the *cloud*. The cloud is cheap, it is reliable. If one fails, its neighbor picks up and keeps working.

At the end of the day, the *cloud* is just a computer on the Internet and you are borrowing it for a task.

What Equipment do I need for my house?

Modem: connects your house to the network

Router: allows multiple devices to talk to the modem

Switch: (Optional) converts 1 Ethernet port to many

Wireless Access Point (WAP): Allows devices to connect using radio (no Ethernet cables)

Gateway: Combines Modem, Router, WAP and Switch into one box.

Modem

The cable that comes into your house will not plug into your computer. You need a translator. For the Cable TV company, it is called a cable modem. For the phone company with twisted pair (pairs of wires) it will be called a DSL modem. If your provider has optical fiber, it will be called an ONT which is short for Optical Network Terminator. Again, with those wacky engineers and their catchy names.

What this does is translate the *network* into
Ethernet. Ethernet is a cable whose connectors
look like big telephone plugs.

Router

Your house, or business, will have one address
on the Internet. Think of it like your mailbox on
the street. But all of us have multiple devices that
want to talk to the Internet. Smart phones,
tablets, TVs and computers all need to be
connected. The digital data that gets stuck in our
one mailbox needs a further address likes Mark's
computer or Lisa's iPad or the living room TV.
The router is the internal mail man that
distributes the digital *mail* (data) to the correct
device.

The device sold to you as a router will include
other functions such as a *firewall*. Properly
configured and kept up to date, a firewall
protects you from much of the evil on the
Internet. The router will also contain a switch.

Switch

The *router* takes care of splitting the modem's
one digital address into many device addresses.
A second device is required to split the hardware
signal into multiple Ethernet ports (connections).
This splitting is called a switch. Most home and

small business routers will include one. You may
need more.

When you need more hardware Ethernet jacks,
you can use a switch. These can add 4, 8, 16,
24, 32 or even more extra ports. The larger
numbers are meant for small offices.

*My home entertainment centers have four or
more devices wanting Ethernet. I have one
Ethernet jack behind the equipment. To get
multiple Ethernet connections I add a switch with
four or eight ports. I mentioned I'm a computer
geek. When I built my home, I ran Ethernet
wires to every room in the house. To allow me to
plug these all into the router, I have a large 32
port switch in my basement.*

Wireless Access Point (WAP)

Most of us don't have Ethernet in every room.
And if we did, we do not want to plug in our
smart phones and tablets or even our laptops.
Most of your devices will want to connect with
wireless or *Wi-Fi*. This is a fancy name for radio
for data. A wireless access point (WAP) plugs
into power and an Ethernet jack. Most often, it is
internal to your router or gateway. You may
want to have more than one in your home or use
something we talk about later called mesh.

Wireless comes in different *flavors* or standards.
The newer standards are backwards compatible
to work with older devices. You really want the

newer standards. Look for *802.11n* or better yet, 802.*11ac.* There is more discussion later in the section on Wi-Fi.

Gateways

Given that everyone needs a modem, router, switch and Wi-Fi, why not put them all in one package? They have and the industry name for this is *gateway*. The down side to the all in one package is that the pieces of technology evolve at different times. Putting them all in one package will typically mean that you have one function that is a step behind the latest technology. But that doesn't mean that it won't work or that you will be unhappy with performance.

Putting it all together.

Your choice of modem is tied to who provides your Internet. You will need a router to connect multiple devices. And getting the signal to those devices requires either a switch or Wi-Fi or both. How you distribute those signals is largely dependent on the size of your home or office and if you have Ethernet wires already installed.

Router Security

A word about gateways and routers: **THIS IS IMPORTANT.** There are bad people out there who want to steal your money or information. In the old days, they threatened you with a gun or knife or hit you over the head with a pipe. Now they sneak in over the computer network. This means that computer security is important. Use good passwords. Change them often and update your software.

Many *low-cost* routers and gateways that you buy on Amazon, at Best Buy, Walmart, Office Depot or other stores can have bad security protection. And the day you take it home from the store, it is out of date. When you buy a network device your first step is to update the software and change the default user name and password. Some cheap products cannot be upgraded. Others are not easy. If you do not know how to upgrade the software in your network devices, you need a different plan.

 I have Apple Wi-Fi access points (WAP). They are expensive compared to Chinese and Taiwanese brands. The Apple WAP can be updated by users without too much hassle. And Apple is great about putting out updates. You can get updates for the higher end companies products like Netgear, but even so, Apple makes it easier for normal people. I've learned recently that Netgear understands the need for helping non-technical people with the upgrades and is working on an application like Apple's program. Maybe by the time you read this it will be released.

Even Apple is not easy. You must load the app and remember passwords. There is an easier option. Get your Gateway (router, Wi-Fi) device from your service provider. (Cable TV company or Telephone company) They lease you the device.

There are often articles in the news where consumer advocates complain about how much more expensive it is to lease a modem or set top from your Service provider. This ignores the benefits of security and software updates that are taken care of for you. And it ignores the technology upgrades that come free with a leased device.

If you do your own IT support, leasing may not be for you. If like many of us, you need a helping hand, leasing is a good option.

Your service provider is highly committed to keeping your network secure. They have engineers that worry about this stuff. They push updates out to your leased device on a regular basis. And you don't have to lift a finger. But you do have to lease the device. It is not that much more expensive and the security is better. Also, if your leased gateway dies, your service provider will give you a new one. If your purchased gateway or router dies, you must go to the store and buy a new one.

*Full Disclosure: In a previous life,
I led teams that designed this
equipment for cable and phone
companies. I currently am a
consultant and some of my
customers are cable companies.
I have strong opinions but I
believe I back them up with fact.
My only connection to Apple is
that I use their products.*

What is Wi-Fi?

Wi-Fi is radio. You use it to transmit data. It can send and receive data to other Wi-Fi capable devices, such as virtually every computer made in recent years, iPhones, iPads and other tablet computers and many video devices. Wi-Fi comes in different flavors. Newer flavors are faster and have more features. These new flavors also are usually backwards compatible with older slower flavors so that they can work with older devices.

There is a professional organization called the Institute of Electrical and Electronic Engineers (IEEE). Among other things they set standards that allow things to work together. The standard that allows Wi-Fi devices to work together is IEEE 802.11. The 'flavors' are letters like 802.11n or 802.11ac.

Old flavors are .11a/b/g. You see them called 802.11g, for example. You want your WAP that is connected to Ethernet to be as new as practical.

In 2017, this means 802.11n and 802.11ac. If you have an older gateway, router or WAP that does not mention 'n' or 'ac' you likely need to upgrade. This is more critical if you are using Wi-Fi for video like You Tube or Netflix or Apple TV.

802.11n and older Wi-Fi flavors use the 2.4 GHz frequency. It can interfere with some cordless phones and other wireless devices. 802.11ac uses 5 GHz and 2.4 GHz. If it says, 'concurrent dual band' then it uses both at the same time. This is important because having both frequency bands is like having two lanes on the highway. More traffic can flow. 5 GHz does not have interference like 2.4 GHz but also does not travel as far. So, you can likely reach more distant rooms with 2.4 GHz. But you get faster data with 5 GHz.

The newest devices offer two 5 GHz pathways. Along with the backwards compatible 2.4 GHz path, these are called "Tri-band".

Wi-Fi access points (WAP) can have different numbers of antennas. 2x2 or 3x3 or 4x4 are common. Bigger numbers are better here. Basically, you get what you pay for within a given brand. Look for numbers in the product name like 600, 1200, 1750, and 1900. The bigger numbers typically refer to the potential speed of the device. Bigger is better in this case.

I said potential. I've noticed that some Wi-Fi connected devices like TVs, and speakers only have 802.11n. This means these are limited to one channel that might be subject to cell phone or microwave interference and will not get the fastest speeds.

The short answer is every three or four years you need to update the Wi-Fi device in your home. And more antennas are better. The number of antennas is not how many are sticking up from the box, but look for the terms 2x2, 3x3, 4x4... This means two, three and four antennas. Antennas sticking up from the box aren't bad, but it is not how you count them.

Mesh

If you are trying to improve Wi-Fi coverage in your home, there are essentially two ways to accomplish this: a more powerful radio or multiple access points. Wi-Fi radios do come in two flavors of power in the US. normal and high power. But even high power is limited by the FCC to avoid interference with other devices. *[If you are reading this in Europe there is only normal power.]* Given this, the best option for better coverage is multiple wireless access points (WAP).

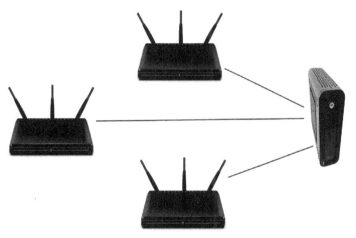

Figure 4 WAPs with physical connection to the router

WAPs need to have a path to the Internet. Some cable modems have MoCA which allows them to send data back and forth over coax cable that may have been installed for cable TV. Or some homes may have Ethernet cables back to the modem. Both options allow you to connect a

normal WAP. See Figure 4 showing a wired
connection, either MoCA or Ethernet, from
multiple WAPs to the cable modem or router. If
you use the same SSID on all WAPs then your
devices will connect to the strongest one. SSID is
the *service set identifier* or name of a Wi-Fi
signal. Older WAPs do not support devices
moving from WAP to WAP. But very often older
homes will have no wired path back to where the
main internet connection is installed. This is
where a different technology called mesh excel.

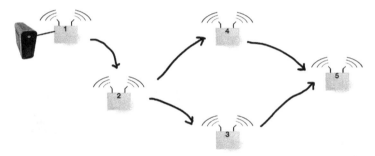

Figure 5 Mesh network

If you don't have wires running to every room,
mesh simply allows WAPs to connect to each
other using the same radios that they use to
connect to devices like your phone, tablet,
computer or TV. See Figure 5. This gives you the
freedom to put a mesh device anywhere there is
a power outlet. The only restriction is that a
mesh device must be in range of another mesh
device. You don't buy one mesh device you get
at least two or more. Data can flow to and from
the master device to a satellite unit and even

from a satellite unit to a second or third satellite unit. Today's mesh units will even support moving devices like a phone or tablet, enabling your device to switch to the strongest signal.

One of the first units to implement mesh was Eero. This product was well received but pricy. Google has entered the market with Google Mesh at a much lower price point. Linksys has Velop. Cable operators are evaluating the technology and will be offering their solutions. The most impressive system I have come across is the Netgear Orbi. The Orbi uses tri-band wireless which includes two channels of 5 GHz 802.11ac. Orbi uses the second channel of 5 GHz to communicate between units and has excellent performance.

I have seen reports testing these units. It shows that using a WAP connected with either MoCA or Ethernet out performs any mesh system. This assumes you have a coax wire or Ethernet wire available. But mesh was invented for those times you don't have a wire. Of the mesh units tested, the tri-band Netgear Orbi shows the best performance. Note that manufacturers are continually updating their devices and this is likely to be a very competitive market.

Mesh devices only replace your previous Wi-Fi devices. They do not replace your router or modem.

Security

Big Sigh!

Recently Yahoo announced 2 different hacks of their security. The first break-in stole 500,000 users' account information. The second break-in stole a million users' account data including passwords, names, addresses and security questions and answers. Yahoo, the second time, forced users to change passwords and security questions. But if you used the same passwords and security questions on other accounts this did not protect them. If you had a Yahoo account in the last five years you need to update your accounts everywhere.

If your accounts support it, use two factor authentication. When you log in from your phone or computer, the site will typically recognize that you are returning. If you try to log in from a different device, it will send a text to your phone or computer and ask you to confirm your identity.

By the way, anything you do to make your security questions hard to guess is good.

If you are reading this book, you are likely not a computer security expert. And if you are like me, you have better things to do with your time. But you cannot ignore it. Let's try and keep it simple.

Passwords:

Never use a word in any dictionary in any language. It will be found by hacking tools. Don't use your name, SSN, birthday or address. My suggestion, take two words like Peach Fuzz, or hound dog, or fast plane, and then pick a pattern to capitalize a letter, throw in a punctuation mark, and add some numbers. eight digits minimum. But longer is better. I've decided in the past month that 8 is not enough. I'm personally moving to longer passwords. You can substitute things that look like letters. *($mart32Fu77 uses the $ for 's' and '7' for 'z'.)*

Be creative, have fun. When you register for something new, they are going to ask you for a password. Set up a system in advance. Try not to use the same one everywhere. You may have the best security in the world at home only to have the company you are doing business with get hacked. You don't have to change much to make the password completely unusable for another account. Say you have an account with REI, take your basic password $mart32Fu77 and add the "R". Now you have $mart32RFu77. For Orvis, it becomes $mart32OFu77. Our basic Smart Fuzz has been transformed and is memorable. Oh, and use a password program to record them and the security questions, just in case. *[I should not need to point out that using Smart Fuzz is now a bad idea because it is in the book.]*

The password program will have its own password. You can sometimes use a thumbprint depending on your phone or computer. But write it down and put it in a drawer, better yet a safe. AND when you update your operating system on your computer, tablet or phone, you need your password there too. And your password program may be not available at this stage of software updates, so write that one down also. Don't stick it in a drawer next to your computer. Move it to the bathroom or bedroom. Unless you are a student living in a dorm or have room mates. Then find someplace better.

Examples of free password programs:

- LastPass
- KeePass
- 1Password
- Dashlane
- RoboForm
- Keeper

Facial Recognition:

Short answer: Not ready for prime time. My buddy got a new computer with Microsoft Windows 10. It has a video camera and the device maker included unlocking the computer by facial recognition. He held his badge photo in front of the camera and the system unlocked. We tried it with various photos and they all unlocked. Do not turn on facial recognition in 2017. Check the website www.simpleguide.tech for updates.

Eventually, facial recognition will use video or multiple cameras for 3-D perception and will be able to tell the difference between a photo and a live face. Cameras are cheap. Dual cameras for authorization should be easy.

Wi-Fi Security:

Older devices came with a default user name of "Admin". The default password was either nothing or "Password". The assumption was that people will change this. But no one did. This is bad. Today, reputable devices come with a sticker on it with a unique name and password. This allows you to do nothing and be *not terrible*. But if you want to be better than *not terrible*, change the user account to your email name or some personal name and create a new password. (See above) **Never** have an *open* Wi-Fi. When you click on the Wi-Fi symbol on your device, the network you are connected to must have the padlock icon. There are several encryption methods. WPA2 is the most common.

Your Wi-Fi signal by default will send out its name. This name is called the *SSID*. You can turn off this feature so that to the casual user your Wi-Fi signal is hidden. But it is not hidden. Smart thieves can still find it. And it makes it harder for other users in your home to find it. Bottom line, use Encryption like WPA2 and good user name/ password choices.

Internet of Things / Home Automation:

In my opinion, Z-Wave security is lacking.
Lacking is a kind word. ZigBee is better, not
great but much better. The AT&T system used
for security is proprietary and is secure. It is also
expensive.

Z-Wave and Zigbee are low power radios used
for Internet of Things applications. They are a
standard which allows different devices to work
with each other.

Wi-Fi, Z-Wave, ZigBee are good for home
automation. But I don't use them for my door
locks. Others might disagree. Some door locks
use Bluetooth and have other features that make
them better. I might consider one of these but I
have not yet.

Credit Cards:

There are two kinds of credit cards. The ones
with traditional magnetic strips give the retailer
all your information. If the retailer's security is
compromised, then all your data is in the hands
of bad people. The newer cards have computer
chips on them. They encrypt your information
and give the retailer a transaction code. This
code is good for the current purchase only. If the
bad guys get access to this code it does them no
good. Your data is safe. Cards with computer
chips, while slower, are more secure.

Apple Pay and other systems:

Apple Pay and other systems are good. Apple pay works a bit like the credit cards with the chips on them. If the store gets hacked, there is only a per transaction code and no account information. Much safer for you.

Chapter 3 Television and Video

Your Next TV will be:

Internet connected and 50 inches or more in size, UHD (4k) with HDR10.

Look for:

HDMI 2.0 or greater, HDCP 2.2 or greater

4 or more HDMI inputs

A Trip through Memory Lane

Back in 1961, NBC started broadcasting "Walt Disney's Wonderful World of Color" on Sunday nights. I begged my parents to buy a color TV, but I was disappointed for another five years. Instead, we would make the pilgrimage to a neighbor's house to view the spectacle of TV in color. The neighbors owned the local

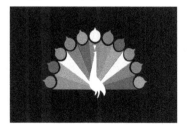

Figure 6 1956 NBC Color Logo Pixgood.com

appliance store and had all the latest technology like color TVs. It seemed to my youthful mind that we were the last people on earth to have color TV. When I found out that some of my friends did not get color until a few years after my family, I stopped feeling like the *poor family* on the street.

Things have come a long way. It turns out that the TV we were watching was *Standard Definition* (SD). Of course, we did not call it that until High Definition (HD) came along. The next big change after color was cable TV or community antenna TV. Due to the advances in satellite broadcasting, we could get channels like Ted Turner's Super Station and later CNN. Around 2000, High Definition (HD) became more commonplace.

> Ted Turner also bought the Atlanta Braves to fill time on his Super Station. Because Channel 17, The Super Station was available everywhere in the US, the Braves became the most popular baseball team in Hawaii.

Then there was (yawn) 3-D complete with funny glasses and headaches for some. Now 4k or Ultra High Definition TV (UHD) is the latest thing.

Storage of TV and movie programming started with video tapes. There was the "Sony vs. everybody else" war (Betamax vs. VHS). If you weren't paying attention, Sony lost. There was the ill-fated laser disc technology. I only knew one guy, Wayne, that had one at home. We had one in the lab to get some early HD content from Japan. But that doesn't mean I spent my hard-earned money on it. After Video tape was DVD then Blue-Ray. Today the rage is *streaming*.

.

Where Video comes from in 2017

Video used to come from one source, your antenna. Today you get video programming from many sources.

Over the Air

Strangely enough, you can still get video from an antenna. It appears people forgot about antennas when cable TV and satellite TV came around, but now that cord cutting is popular, people are remembering that TV stations still broadcast TV over radio waves that can be received for free. The primary broadcast stations converted to digital *high definition* (HD) by 2009. If you had an older TV you needed a converter box to receive the new signals.

Today, new TVs have tuners that will pick up today's broadcast TV. The signals are digital. This allows the computer chips in the TV to ignore most noise. If you have more noise than the computer chips can deal with you will start to see little squares in the picture. If the signal gets worse, you get no picture at all. Digital looks great and then suddenly stops working. If you have this problem, you need a better antenna or an amplifier. To find what you need you can search for "HDTV antenna selection" or try a site like www.antennaweb.com or www.tvfool.com.

I put an antenna in my attic. I had to add an amplifier when we put on a new roof. I have a stronger antenna on the *outside* of a vacation cabin 60 miles north of Atlanta and I get good signals.

You can feed the signals straight into your TV or you can buy a digital video recorder (DVR) like TiVo. You will use coax cable to carry the signal. A device called a splitter will allow you to serve multiple TVs. Be careful with splitters, each additional output lowers the strength of the signal. You might need to add an amplifier to compensate.

Figure 7 Splitter

I get 60 channels from my local broadcasters. Some are music channels from local radio stations. Some are in Spanish. Some are religious. There are 6 to 8 mainline commercial HD stations that I get for free. The signal is often better than the one I get from cable or satellite.

Each local station typically carries the main programming in HD. They also offer other auxiliary stations that are often in standard definition (SD). These are often old reruns from decades ago or the local weather information. Take for example channel 2. The main program is in HD and will be listed on your TV or DVR as channel 2-1. The auxiliary programs will be 2-2, 2-3, 2-4 and so on.

Just when we get this figured out, changes are coming. The federal government wants to free up space in the radio spectrum to re-sell to cell phone companies. The extra radio waves will be used to improve connection speeds for smart phones. Some TV stations may sell their channel back to the government and either only be available through cable or satellite or they will rent space on another station's channel. For example: Channel 46 could sell its channel back to Uncle Sam and rent space on Channel 2. You would still see Channel 2 and Channel 46 but channels 2-2, 2-3 and so on would likely disappear.

The Channel number is no longer associated with the frequency of the signal like it was in the early days of TV. Today the Channel number is just a brand or name. To find your channels, your TV must scan the air waves to find TV programming. It picks the name or channel number out of the digital signal.

The other change coming is ATSC 3.0. ATSC is simply the Advanced Television Systems Committee. ATSC replaced another bunch of initials, NTSC, when we went digital. The new standard, 3.0, is not backwards compatible with your current TV. When it comes, you will need a box or a new TV. South Korea is switching this year. The US is testing in North Carolina. No date for the change has been announced. But you can expect the transition to take a long time. The last transition from NTSC to ATSC 1.0 took ten years. The new standard completes the move to internet protocol and will offer cool features and more services. At least, that is the sales pitch.

Cable and Satellite

Cable and satellite are called multi-channel providers. They offer hundreds of channels including local off-air broadcasters, premium channels like HBO and Video-on-Demand (VOD). The largest cable providers are Comcast, Charter and Cox. Satellite providers are Dish and AT&T Direct TV.

Cable requires that you have a coax cable or fiber optic cable physically connecting your house to the network. This cable can offer data and phone service as well as video.

Both AT&T and Verizon have growing fiber optic networks and offer some of the same services on those networks as the cable operators.

Satellite requires a dish antenna pointed at the sky and a box connected to your TV. Satellite works well for video broadcast but not for data or phone service.

Cable networks used to carry the old analog TV signals for many channels even after the broadcasters converted to digital. An older cable-ready TV could connect directly to these channels and view them directly. Today, most cable operators are moving to all digital. If you have an older TV you need a box to allow you to view programming. This is a different box than the one that allowed you to get digital programming from an antenna. The cable version is called a digital terminal adaptor (DTA) or sometimes mini-box. These are inexpensive but do not have many features like the guide or access to premium channels or video-on-demand (VOD).

There is one other item we need to discuss for cable; CableCARD. Cable TV set tops, the box the cable provider leases to you, do many things. They tune the signal, they provide guide data, the have extra features like video-on-demand (VOD), a digital video recorder and they de-crypt the signal. Most services on the cable are encrypted and are only available to paying subscribers. This requires a box from your cable company to decrypt the signal.

Television manufacturing is a highly competitive business. TV manufacturers make their money not on the display, but on the extra features. Before internet connected TVs, having a set top box between the TV and the cable made the extra features not worth the expense. TV manufacturers went to the government asking for a solution and the CableCARD was born. This moved security from the set top box to a card you insert into your TV, your TiVo box or a cable company set top. This card could be used with any TV that had a slot and passed certification by an organization known as Cable Labs. And now that they had the ability to better compete, the TV companies didn't. And the whole thing has died a slow painful death. It did allow TiVo to build an alternative set top DVR. And it allowed traditional set top vendors to sell their products into competing networks. But the whole TV thing was stillborn. There is a lesson in there somewhere.

Blu-Ray and DVD

More and more content is streamed over the Internet. However, if your kids or grandkids want to watch Frozen again and again, having a copy of it on physical media can be nice. Enter Blu-ray and DVD.

DVD came first. It was not full HD but you could get better resolution than SD.

Blu-ray has more capacity. Basically, it holds more bits. Blu-ray disks are all HD and some newer ones are UHD (4k). *[We talk more about UHD a few pages from now.]*

Both disk formats carry extra audio information for better sound. And they may have additional features like extra video or the director's comments.

Internet

Fast internet access has enabled both audio and video streaming. You will hear the term *broadband* tossed about. Broadband has a specific legal meaning defined by the FCC; it is internet service with speeds of 25 Mbps or greater. That is 25 million bits per second or more. This speed is almost never available on DSL from the phone company. It is available on fiber optic services and cable modem services from the phone and cable companies. You can get video services over DSL. However, the number of simultaneous channels are limited and UHD (4k) may not work. When we refer to fast Internet, we are talking broadband speeds of 25 Mbps or greater.

Video over the Internet is often called an Over-the-Top (OTT) service. The companies that lease your internet access also sell video. When you use the Internet to get video rather than buying their video service, you have gone over-the-top. This is legal and it is competition at its finest.

Video on the Internet is provided by companies such as Netflix, YouTube, Amazon, Sony, Hulu, Direct TV Now and many others. Major services like HBO and ESPN are starting to offer content direct to their Internet only customers. Cisco has reported that Netflix is 60% of the traffic on the Internet during prime time. Many cable operators report that YouTube is the second biggest user of Internet data after Netflix. This market is changing rapidly and this paragraph will be out of date before the ink dries. See the blog www.simpleguide.tech for updates.

You need a box to get these video services.

TiVo and Cable TV Set Tops

TiVo now uses the Internet to get data for its guide. In addition to tuning the off air or cable signals, TiVo offers a collection of apps to get content from the more popular video sites like Netflix and YouTube.

Cable operators in trying to be more competitive are starting to view some services such as Netflix as an enhancement rather than a competitor. Check with your cable operator to see which ones are available.

Apple TV

Apple is on its fourth-generation TV box. It is the only platform for iTunes content. It does not offer UHD(4k) content. On the fourth-generation box you download the apps you need to access your favorite content. It does offer Siri and a microphone that allows you to navigate by voice. It also uses Apple Air Play to allow you to easily show content from your Apple computer, iPad and iPhone on the TV.

Roku

Roku does not offer any of its own content, and does not compete with any of the content providers. This allows it to offer the widest variety of applications and content *[But not Apple iTunes]*. Comcast cable even has an app to allow you to watch Comcast content on their network using your Roku. If you are using the Comcast app, you still need to pay a monthly fee for the guide data. And you need to have Comcast video service.

Roku also has a remote with a microphone to allow voice navigation. The Roku Premier + and Roku Ultra have UHD (4k) capable interface with HDR10. To view UDH content, your TV must be UHD compatible and have HDMI 2.0 or greater and HDCP 2.2 or greater. *[We discuss these two standards later in this chapter.]*

Amazon

In addition to having the Amazon Prime video service, Amazon offers different hardware for accessing video over the Internet. The Amazon Fire is an internet connected set top box. Amazon also offers a smaller Amazon Fire Stick which plugs into the HDMI connector on most TVs. The remotes offer Alexa voice control.

Google

Google offers the Chromecast series of internet connected hardware. Google has a music Chromecast. Google also has the original Chromecast for video and the newer Chromecast Ultra. The Ultra offers UHD (4k) video in HDR. Chromecast units plug into the HDMI port of your TV. An app on your smartphone is used as the remote.

Apps on the TV

Smart TVs now come with apps. Many of the same apps that you get on the other platforms are now directly available on your smart TV. This can be good or it can make your life harder. If your smart TV has the apps you need, you don't need a separate box. Many apps are available only one select boxes.

Having apps on your TV means that you now have a menu. Technically you always had a menu somewhere but it was hidden from daily use and you could ignore it until you wanted to adjust something. On some smart TVs, you must first go through the menu before you can watch anything including broadcast TV. And some smart TVs come with imposing remotes that only the most hardcore technophile could love. When you buy a smart TV, look out for these issues and make sure you can use it before you take it home.

TVs connected to the internet have other pros and cons. One advantage, beside directly accessing internet video, is the ability to upgrade the software in the TV. Yes, dear reader, it has come to this. Your TV is now a computer in disguise. It has software. You must now keep your TV software updated to avoid hacking just like your computer, phone and tablet. The advantage is that you can also get a fix to problems. I have seen HDMI connection issues resolved with a software update to the TV. The down side to TV internet connectivity is, of course, hacking and some unscrupulous behavior by some low-end brands. Vizio TV was taken to court over collecting user viewing data and selling the data to marketers. Vizio claims it was not user identifiable and security experts disagree with that assessment.

Choosing a TV

We have talked a lot about what to watch, but we promised to explain the latest buzzwords and give you advice on what to buy.

HD or UHD

High Definition (HD) TV was a big step up. The extra resolution in the video was noticeable. Going digital helped make the pictures sharper and clearer. The funny story in my home was during the Winter Olympics games. My wife was at the neighbor's house watching ice skating. She came home saying that on Vanessa's TV she could see the sequins on the skater's dresses and she wanted that. What guy on earth would resist the invitation from their wife to go buy a new larger TV? I seem to recall we went out shopping that very night.

Ultra-High Definition (UHD) is also called 4k and has 4 times the number of picture elements (pixels) as HD TV. It has tons more detail. But there is a problem. The limitation is with the human eyeball. Your eye can resolve about 2 degrees. Any less than that the sensor in the average eye cannot tell the difference. When you are about a foot away from an UHD screen in the store, it looks great. But when you step back far enough. You don't see the extra pixels. The average distance between where people sit when watching TV and the set is nine feet. To see all the extra detail in UHD the TV at nine feet needs to be 120 inches.

Does this mean you should not buy a UHD TV? No. There are two reasons behind the "No." The first reason is that building a UHD TV costs only about $20 to $40 more than HD. That cost is dropping. What you will see in stores is that any TV larger than 40" or 50" will only be available in UHD. Resistance is futile. Get a bigger set or sit closer or simply enjoy having the coolest TV on your block. (Tom, this is for you.)

Does this mean that there is no benefit to UHD? Again, No. UHD comes with another feature, HDR or High Dynamic Range color. Specifically, it comes with HDR10. Color in the world of HD was eight bits. Some UHD content will still be eight bits. By content I mean movies and TV shows. The newer, better UHD content will be ten bits. Every time you add a bit, you double the information. Ten bits gives four times the color information. With a 60" TV at 9 feet in a blind test, consumers were shown HD and UHD content both in eight-bit color. The consumers could not tell the difference between HD and UHD. Then the testers turned on HDR10 and the consumers went "Wow!". The difference was immediately apparent. The hidden jewel of UHD may be better color. (and the justification for another set. I'm waiting for the next Olympics.)

If ten bits is better than eight, what would twelve look like? The good folks at Dolby Labs have a twelve-bit standard. But they must employ marketing people and not just engineers. They don't call their system HDR12, rather, they call it Dolby Vision. TVs with Dolby vision cost more to build and they will cost you about $20 more than sets with HDR10. In 2017, you are likely to find Dolby Vision only on the best TVs in a manufacturer's product line. Not all movies and TV shows will be recorded or broadcast in Dolby Vision. But the ones that are will look noticeably better. The TVs will handle the differences between all the color standards. Regardless of the number of bits used to show the color, TVs will do the right thing. If a TV only has HDR10 and the movie is recorded with Dolby Vision, the movie will appear in HDR10.

LED/LCD or OLED

LED/LCD TVs are larger and cheaper

OLED TVs have better colors and deeper blacks, they have better contrast. They have a wider viewing angle

Both have the same color range and resolution. Similar energy and lifespan

The original TVs were cathode-ray-tube (CRT). CRTs sound like something out of Buck Rogers. They used "ray-guns" to draw a picture on the screen. They have largely become extinct. When HD flat panel TVs first came out, the choice was LED/LCD or plasma. Today LED/LCD TVs have gotten so good that you don't find many plasma TVs. They had shorter life and used more power. Today there is a new technology called OLED. Let's talk for a minute about the differences and why one might be better.

LED or LCD is the same thing. The light produced by the TV is from an LED backlight. LED stands for Light Emitting Diode. These TVs do not have an LED for every pixel. Rather they use LEDs on the edge of a material to produce a uniform back light. For every picture element (pixel) there is a liquid crystal element. These LCDs basically act as gates to the light from the back light. They have colors. To get a certain color pixel, the gates allow different amounts of light through the red, green and blue filters.

OLED is a recent development. It stands for Organic Light Emitting Diode. There is no backlight. Every pixel has a red, blue and green LED. It is easier to get black because you simply don't generate light at that pixel. You are not trying to block the light like LED/LCD. The contrast ratio is higher because of this. These panels are thinner because there is no backlight plane. Some smaller ones are even flexible. Apple is rumored to be using OLED on the new iPhone 8. It is easier to see the TV when viewed from a side angle. There are currently fewer manufactures capable of building large OLED panels. With lack of competition and greater expense to build, these TVs cost more. However, at the high end, the cost of the best LED/LCD TVs are comparable to OLED. If you are buying at the high end, look hard at OLED.

Inputs

UHD requires HDMI 2.0 or higher.

For now, I am going to ignore older connectors. All video in 2017 is carried on HDMI. You might have an old video game with an older connection. But current video is HDMI.

HDMI is High Definition Media Interface. It carries digital video and digital audio for multi-channel sound. Some later versions carry communication between the TV and the attached equipment. The advantage is there is one cable and connections are simpler.

Figure 8 HDMI type A Wikipedia.org

HDMI cables are fragile and are sensitive to dust. I'd cap or cover unused connections to keep dust out. I also recommend supporting the cable. The weight of an HDMI cable can put stress on the TV connector and in some cases, cause it to fail sooner than you would like.

A few years back, my buddy, Tom, wanted to buy a huge UHD TV for his new home. My recommendation at that time was to wait. UHD comes with new versions of HDMI and the associated copy protection. At that time, you could buy a UHD TV but it still had the older more limited connections. Really? In what universe did that make sense.

UHD needs a version of HDMI that operates at a higher speed than HD. It needs HDMI 2.0 or higher. You can get some lesser quality UHD with HDMI 1.4a, but that is not what you are buying UHD for. This is not a problem in the big-name brand current model year TVs. If you are looking at the bottom end of the cost spectrum you need to be careful, even in 2017

Copy Protection

```
UHD requires HDCP 2.2 or greater
```

Along with HDMI comes HDCP, High Definition Copy Protection. HDCP is not always present. HDCP is not used for YouTube videos and other user created content. HDCP is used to protect commercial content such as TV shows or movies. HDCP encrypts content on your HDMI cable.

When content went digital, it was much easier to make perfect copies which can then be pirated. The people that make TV shows and movies insisted that their efforts be protected by something called copy protection. For HD over HDMI 1.x, the copy protection used is HDCP 1.3 It has since been defeated.

Because HDCP 1.3 has already been broken, people creating UHD TV shows and movies wanted a better system and HDCP 2.2 was introduced. If you are going to watch UHD content, your TV must have HDCP 2.2 or greater. Again, for well-known TV brands this is not usually a problem since 2015. Some 4k computer monitors have HDMI 2.0 but do not have HDCP 2.2. Some low-end TV brands do not have HDCP2.2. If you plug your Roku Ultra or equivalent device in to a UHD TV without HDCP 2.2 you get a message saying you cannot watch the shows in UHD.

It was announced recently that HDCP 2.2 has been broken and the companies involved are being sued by the movie companies. But that doesn't eliminate your need to have this standard on your new UHD TV.

Other Considerations

When buying a TV, I would suggest getting one with 4 HDMI connections. Some TVs have 3 or less. This may work for you but I think 4 is a better number.

Higher contrast ratios are better. The contrast ratio is the difference between light and dark. But, there is no standard test for contrast ratio. You cannot compare published numbers on a Sony to a Samsung or LG. You may be able to compare one Samsung to another Samsung. This assumes all groups in Samsung use the same test which is likely, but not required.

Connecting the pieces together

When you had only an antenna and two or three channels, connecting the TV was simple. We have discussed all the different sources of video available today. The trick is to select the input and then juggle remotes to control the system. In the limited space we have here, let's talk about the simplest case with all sources using HDMI. This is a reasonable assumption. HDMI carries both video and audio. You only have one cable to worry about.

I'll talk about strategies for dealing with remotes in a few pages. If you need more advice on a different setup, use the contact page on www.simpleguide.tech or look for "The Simple Guide to TV" due in print in late 2017.

Unless you are using a home theater system, let the TV switch between audio sources. If you use the TV speakers, no extra work is required. If you use an external speaker but not a home theater system, such as a sound bar, [*more on this later*] take the audio output from the TV and plug it into the external speaker.

Figure 9 Optical Digital Audio

It will need to be an amplified speaker, like a sound bar. The best connector to use is the digital audio output assuming your sound bar has this as an input. Most of them today are optical. Using the digital audio output will give you clean multi-channel sound.

TV with Antenna

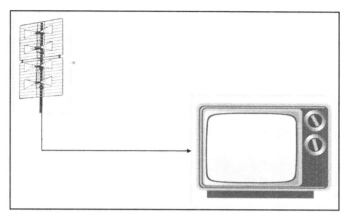

Figure 10 TV with Antenna

This one is straight forward. Get a new antenna that is designed for today's digital signals. See the web resources listed earlier for recommendations based on your address. You will need to run coax cable to the TV.

Your TV needs to have an *ATSC* tuner. Most digital TVs made since 2012 should have this. Connect the two together and open the menu of the TV. Have it scan for channels. Then go through the channel list and delete all the junk channels you will never watch. You will thank me for this step.

Congratulations, you are done. It used to be that Channel 2 was on 55.25 MHz. This is no longer true. The TV will find the channels by looking at all the possibilities. The name of the station such as "Channel 2" is carried in the signal and this is how it will show up on the menu. Remember the auxiliary programs on 2-2, 2-3 and so on.

TV with multiple Inputs

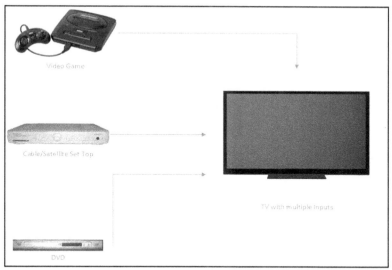

Figure 11 TV with multiple inputs

If you have an input on the TV for all your devices, you can simply plug them in and select your source of video using the TV remote. You can also use the TV to select and watch a program from the antenna.

If you use an external speaker such as a sound bar, connect the sound bar to the digital audio output of the TV and your audio and video will stay in sync.

TV with Home Theater

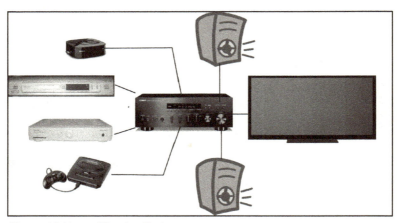

Figure 12 Home Theater System

I hope with a home theater system you have at least six speakers to get the whole effect. The figure above shows two for simplicity.

A home theater system will give you many options and with those options comes some complexity. Your old stereo system allowed you to choose the source of your music such as radio, turntable, CD player and play the music on speakers. A home theater system adds video to the audio. You can still have audio sources like radio connected to the system but now you can also have video. To add video the home theater system will have HDMI inputs that carry the video and the audio from digital video recorders, IPTV boxes like Apple TV, and game systems. The home theater receiver will decode and amplify multi-channel sound, typically 5.1 or 7.1

and control which source of video is being fed to the TV.

Complexity #1: Even if you are using the home theater equipment to select video source, if you have your off-air antenna connected directly to the TV rather than connecting to a set top box such as a TiVo DVR then the TV is used to choose the source of the video. If the TV has apps like Netflix, YouTube or Pandora, then once again the TV is the source. To get the audio from the TV to your nice multi-channel speakers you need to take the digital audio output from your TV and connect it back to the home theater amplifier. This is not shown on figure 12.

If you are watching video that comes from the TV, first select the content on the TV. Second you select the TV as the source of the audio on the home theater.

If you are watching video that is connected to the home theater, you select the source on the home theater and then on the TV select the input that is connected to the home theater.

I have not seen a case where the TV has content that is not available on a box that can be plugged into the home theater. You may be required to buy an extra box but it could be easier for you to operate.

Complexity #2 UHD. We said earlier that UHD requires HDMI 2.0. If you have owned a home theater for several years you might find that it only has HDMI 1.3 or 1.4. This means either a new Home theater amplifier or you need to do something different. In truth, the speakers are typically the most expensive part of the home theater system. In 2017, a good 7.1 channel amplifier is less than $1,000.00. I bought a nice Sony for a bit over $200. Upgrading your amplifier to handle UHD with HDMI 2.0 inputs and outputs may be the easy approach.

If your new UHD TV has enough inputs there is another path. You can feed all your UHD HDMI 2.0 devices into your TV and connect the digital audio back into your home theater. Digital audio is the same. There is no version number to contend with.

Complexity #3: Dolby Atmos audio. The good folks at Dolby have found a way to improve the sound in your home theater by adding speakers in the ceiling. The details are covered in the chapter on audio. What is important to note here is that Atmos adds two to four additional channels and a different audio processor. Bottom line, you need a new amp to get Atmos. Many UHD Blu-Ray discs have a second audio track with Atmos. Fine, if you want it, but the disc will also have 5.1 or 7.1 sound tracks. Regardless of which sound track you use, your digital audio connector will work.

What to do with too few inputs

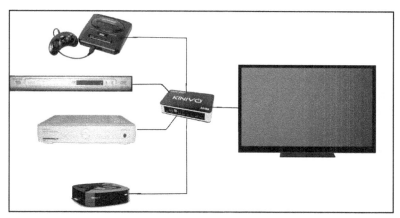

Figure 13 TV with too few HDMI inputs

I have an older HD TV in the family room. It only has one HDMI input. I don't have a home theater system to switch video inputs. I need plan B. Fortunately several companies make an HDMI switcher. Look on Amazon for "HDMI Switcher". There are several brands. I show a Kinivo in my diagram. I've used several at work and have one at home. They work. Working is a nice feature. You use the Kinivo remote to select your source. You use the remote that came with your source to select and play the video and you will use the TV remote to control volume. Unless you have a programmable or Universal remote. *[See the next paragraph.]*

If you want better sound, connect the TV's digital audio output to a sound bar.

The Problem with remotes

As you start to see from the preceding paragraph, all the bountiful sources of video come with a plethora of devices each with their own remote. The classic solution is a basket to hold them all. Let's talk about why there are so many.

There are only two or three chip vendors that make remote control chips. It shouldn't be that hard but it is. Each manufacturer of consumer electronics has their own control language. Often, they have many *dialects*. Your Sony TV may use Sony 34 and your Sony Blu-Ray player may use Sony 98. I don't know why. Usually, both dialects work except for some small thing. You would not use a Sony code to talk to Panasonic.

The chip vendors have big libraries of codes and know how to talk to which devices. I don't know why there is not a standard except that once you buy brand X TV, it is easier for you to use brand X for all the other peripherals.

Programmable Remotes

The remote you get with your TV or your cable set top or satellite set top will often have buttons to allow you to teach the remote how to talk to the other devices in your setup. Your installer may set it up for you or it may come with a little tiny manual written in something pretending to be English. Let's give an example. But remember every device is different. If you are trying to program your cable company remote to control your TV you need to use the little manual to know what buttons to push. If you don't have the little manual, you need to Google, "How to program ..." and name the company or remote. You can also call customer service.

The instructions will be like these for the Charter cable remote that I found on ebay.

- Get familiar with the remote. If you delay too long between button presses, the system times out and you need to start over.
- Each device you want to program will have a list of codes. Get them from the little manual or the web and write them down for easy reference.
- This example assumes you are programming the Charter remote for a TV.
  - Using the TV remote, turn on the TV.
  - Press the TV button on the Charter remote. Ensure the TV button on the Charter remote blinks once.
  - On the Charter remote press setup. (some remotes use a two-button sequence.)
  - Quickly enter the 5-digit code from the top of the list of codes for the TV.

- o Aim the Charter remote at the TV and press the power button.
- o If the TV turns off, you are done.
- o If the TV does not turn off, repeat the steps with the next code on the list. Repeat until the TV turns off.
- Repeat for the other devices in the system.

Some remotes have a "911" feature. This allows you to let the remote find the correct code. It is time consuming. Other remotes have a learning feature. You put your universal remote into learning mode, select the control you want to teach and point the remote from the device at the programmable remote and issue the command using the device remote.

Assuming you have successfully programed your programmable remote, you will run into the problem with modes. Let's explain this in terms of languages. Your TV speaks Sony. Your set top speaks Motorola. Your DVR speaks Panasonic. Your remote has been taught to speak Sony, Motorola and Panasonic. It just doesn't know when to speak in which language. Before you push a button, you must tell it which device you want to command. You cannot turn the TV on if the remote is in DVR mode. You cannot fast-forward the DVD if you are in set top mode. You control these modes with the buttons that name the device. To be fair, some buttons like volume are always associated with the TV. Which works unless you have home theater.

I said earlier that HDMI allows devices to talk to each other. If you are connecting all your devices to the TV, some high-end TVs can figure out which devices are connected to which inputs. Selecting a source sets you up for success using the TV's remote. At least that is the theory.

Universal Remotes

Universal remotes can control more devices than a simple programmable remote. Some of them can be tied into home automation systems and draw the curtains and dim the lights as well as control the entertainment system. They avoid the mode problem by being able to talk multiple languages all at the same time. They do this by allowing you to define *scenes*.

A scene is a high-level task description such as watching Netflix. You program the scenes on your computer using a program you download from the Internet. You tell the program which devices you have and what scenes you want to create and it will walk you through the rest of the process by asking questions. It is all pretty simple. In the end, you connect the remote to the USB port on your computer and it programs it. No multiple key presses. No 5 digit codes. No mode buttons.

In my system, to watch Netflix, I turn on the TV, switch the input to the Apple TV box and turn the Apple TV box on. I then use the navigation and select key on the remote to navigate through the Apple TV menus.

For other sources of content, you create different scenes. I have a scene for off-air and a scene for Roku.

The down side to Universal remotes is timing. The remote tells the TV to turn on and switch to input two for the Apple TV box. The TV is busy turning on and doesn't listen to the second command. There are many timing situations where without feedback, the system seems to get lost. I've had good success with a Logitech Harmony One. But there are other brands out there.

My prediction is that as devices add Bluetooth radios, the remote will move to an app on your phone. This allows feedback for the remote and eliminates the need for a computer to program the remote.

Improving the sound

There is a whole chapter on sound. I'll cover many details there.

When you go to a theater, you get the whole experience. Big screen immersive video and big sound. Many people are buying bigger and bigger screens for home viewing but they are skimping on the sound. Combine this with the trend for TVs to get thinner and thinner and it is unreasonable to expect good sound from the TV. You need something better.

Home Theater

Home theater adds multi-channel sound for that immersive sound experience. A typical system is 5.1; two front speakers like a stereo, a center speaker placed with the TV for dialog so that speakers on screen sound like they are on screen, and left and right rear speakers for that surround sound effect. The ".1" is the sub-woofer. This is a speaker for deep bass like explosions or the deep rumbling of an earthquake. Low notes produced by the sub-woofer are not directional. You can place a sub-woofer anywhere in the room.

Some systems add rear speakers for a sixth or seventh channel. The story goes that these were added so that when space ships in Star Wars flew overhead, the sound came directly from behind.

Dolby Atmos adds two to four speakers for a vertical dimension.

Sound Bar

When you have a home theater with 8 speakers it requires lots of wires (unless you use wireless like Sonos) and takes up space in the room. An easier solution is a sound bar. Sound bars use some computer magic and will bounce sound off the rear wall to give you a good imitation of a home theater system with one box. You can even add a sub-woofer for extra bass. Sound bars are a compromise but sound much better than the TV.

Chapter 4 Telephone

Bottom Line: The easy solution is a cell phone if you have good reception in your home.

The lowest cost is VoIP (Voice over Internet Protocol)

Traditional

The first telephone patent in the US was granted in 1876 to Alexander Graham Bell. The classic phone has improved, but it has not changed much. In the beginning phones were connected directly to each other. This obvious limitation was overcome by switching centers that allowed you to use a single phone to talk to several different people, one at a time.

Figure 14 Candlestick Phone Pixabay.com

Classic telephone is made possible by two wires. This two-wire, twisted pair solution is referred to as POTS, "Plain Old Telephone Service." Because it uses physical wires it is also called a *land line*. Once upon a time, it was provided by a giant monolith called the Bell System. The courts declared that this was a monopoly and broke up the Bell System into many smaller companies. Most have since rejoined to form AT&T and

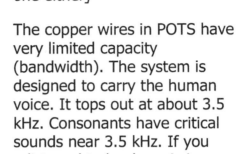

Verizon. *[I can't figure out that one either]*

The copper wires in POTS have very limited capacity (bandwidth). The system is designed to carry the human voice. It tops out at about 3.5 kHz. Consonants have critical sounds near 3.5 kHz. If you

Figure 15 Data Modem   listened to loud music in your youth, you may have trouble hearing these. Your spouse is right; you are hard of hearing.

Sending data over such a limited medium is difficult. The first data modems (modulator/demodulator) were limited to 300 baud, roughly 300 letters per second. They have gotten much faster since then, and the phone company now takes advantage of all that old copper in the ground to use multiple pairs of wires to send data over DSL (Digital Subscriber Line). There are various forms, VDSL and ADSL, but it is pretty much the same. The amount of data you get depends on the distance from the phone company's equipment. Six to twelve Mbps is typical.

The nice thing about POTS is that it is battery powered. There are large batteries at the phone company and having service is not dependent on having power in your home. And because people use the phone to call police or ambulances, government has decided it is a *life line service.* If you can't afford it, the government will even subsidize your phone.

No surprise, telephone service like everything else is changing. Consumer demand for faster data for the Internet and even video programming is driving the need for higher capacity. These capacities used to be 300, 1200, 9600 or 56K baud. Now they are listed as 6, 12, 25, 50 Mbps (Megabits per second or million bits per second) The FCC has said that a connection must be at least 25 Mbps to be called *broadband.* This pretty much rules out most DSL services for the average person. And there are other reasons for the phone network to change.

Because copper wires rust (oxidize, corrode if you must) especially if located near the ocean, and because the capacity (data speed) is limited, phone companies are moving to fiber optics. This is glass or plastic wires the thickness of a human hair that have oodles of capacity and don't rust. But the battery back in the main office cannot power the box in the subscriber's home. The phone doesn't work if your home electricity is off. So, the government has said that you must be able to purchase, at your expense, with no price limit, a battery for your fiber optic box (ONI, Optical Network Interface or ONT Optical Network Terminator) to keep the phone alive for

8 hours. [In February 2019, this moves to 24 hours] The battery must only keep ring tone working: no data, no long conversations.

Landline usage peaked worldwide in 2000. Since then the number of landlines in use has been slowly declining. In 2000, I had three landlines. One for the data modem, one for the kids and one for the parents. By 2002, we had shifted to Cable Modems and eventually the kids grew up and left home. Eventually we moved our remaining landline to VoIP which I will describe later. My kids, when they got their own homes, never got a land-line (a phone connected to physical wires). Instead they rely completely on cell phones. This trend is moving up the age brackets as speeds and coverage improve and as prices rise on land lines.

Wireless or 'Mobile' (Cell Phone)

Cell phones are so named because they use a wireless network divided up into *cells*. The phone uses radio to transmit to the strongest cell and if the user is moving and another cell becomes stronger,

Figure 16 Cell Phone Pixabay.com

the call is handed off to the cell with the stronger signal. Signal strength is indicated by the number of bars on your phone display. This hand-off usually happens seamlessly with no interruption in the call. Because the radio frequencies are shared by many cells, the system can handle a great many calls. Because the devices are mobile, they have batteries and inherently satisfy life line service requirements. (Remembering to charge them is your job.)

The first cell phone was demonstrated by

Motorola in 1973. The first time you could buy one was in 1983. Cell service is now available in every country with seven billion devices in use. More people have cell phones than electricity.

Motorola is no longer in that business. The top manufacturers are Samsung(Korea), Apple(US) and Huawei(China).

Figure 17 Original Motorola Handheld Wikipedia.org

Note: this maybe the only phone your children and younger co-workers have.

Cable TV

The cable TV providers have a coaxial network. This is basically a more expensive wire compared to twisted pair that is much better suited to carrying

Figure 18 Coaxial Cable Wikipedia.org

lots of TV channels over long distances. If the phone company bandwidth ended at 3.5 kHz, the cable TV cable was capable of 860 MHz or more. The cable TV guys figured out that the network was good for more than just video; they could use it to send digital data. If you can send digital data in both directions, you can have a phone call.

The first widely available digital service on cable was Digital Music Express (DMX), a subscription music service. This was followed by SEGA Channel for delivery of SEGA Genesis video games over the Cable TV network. These were both download only. The DataXcellerator Cable Modem allowed for >5Mbps downstream but used a telephone line for reverse communications. Cable modems, using cable for both directions, came along a few years later.

The phone portion of a cable modem is called an eMTA (embedded multimedia terminal adaptor; just rolls off the tongue, doesn't it?). Essentially it is Voice over IP (Internet Protocol). The eMTA turns your voice into digital bits and the cable modem sends it up and back over the cable system. There is a lot of software involved, and the cable provider must connect you to the regular phone system. But it is simple on your end. Just plug your land line phone into the unit.

Like fiber optic systems, the power for your eMTA

comes from your wall plug. If you want to make a call when the power is out, you must have a battery. Your cable TV service provider is required to offer you the option of a battery at your expense. As of 2016, this battery can be capable of eight hours of being available if you request it and pay for it. In 2019 that time increases to 24 hours. But then everyone has a cell phone, right? Why pay for an eMTA battery if you have a cell phone?

Figure 19 eMTA

Your standard phone can simply plug into your eMTA. Initially, some alarm systems had difficulties but my experience is that these problems have been resolved. Basic eMTAs have two phone connections. Models with four, eight or more are available for small office configurations.

When power goes out, your eMTA will complete a call if you are already using the phone. It will then reboot into low power mode to allow the battery to last the required time. This provides phone service but no Internet. When power is restored, the device will reboot when the phone is not busy and restore internet services.

Internet

If you have Internet from whatever source, you can have a third-party VoIP solution. VoIP is Voice over IP or internet telephone. Examples of VoIP solutions are Magic Jack, OBi, Ooma, Vonage and others. Be sure to check for current reviews, as service can vary over time. What was good for your neighbor last year may not be good this year. This is a fast-changing market.

I bought a device that provides the service from a local big box store. I plugged it into my computer via USB. (If you are a tablet-only house, this will not work.) I logged into the web page and set up my account. They gave me a temporary number and offered to have my original home phone number transferred. They charge an extra fee for this, but it was small. Do not cancel your phone service before doing this. (Your number will disappear forever) I got an email in a week or so saying my number was now working on my new VoIP device. I paid about $60 for the device and a year of service. I paid $20 to transfer my number. Every year I pay another $30 or so dollars for unlimited calls to the US and Canada. This allows me to call my sister in Toronto without breaking the bank. You pre-pay for International calls but for major countries the cost is pennies per minute. I did have about $8 left over on my international calling plan that disappeared at the end of the year. (ooops; I did not know that would happen) Whatever your locality charges for 911 service will be passed along to you. The 911 tax in my area is about $15 per year. Looking at reviews of the service I selected, people either love it or hate it.

Choosing your poison

If you are reading this book chances are you are not a technical wizard. Things that are easy and obvious to folks in the industry are not easy and obvious to my neighbors and my parents. I tried to compare choices below, but as I created the table, it is not so simple.

If you don't have a computer, don't try to do third party VoIP. Stick with a Service Provider (classic, cable or wireless). VoIP is the cheapest option by far. It does help to know how to connect Internet and phone hardware. These products are designed for the average person. But you need a computer and you need to know how to do things on the web. If that scares you, and maximum saving is not what you are about, then keep looking.

Most younger folks only have a wireless phone (cell phone). Maybe one day we will get a phone number at birth, and it will follow us to the grave.

My daughter's number still reflects the city where she had her first job. I guess no one pays for long distance anymore.

Comparison of phone choices

Feature	Cost	Ease of setup	Voice Quality	Long Distance	Power Outage
VoIP	Low	Medium to High	Good to OK	Cheap	No phone
Wireless	High	Easy	Depends	Based on Plan	Battery included
Cable	Medium	Easy	Good	Typically, lower	Battery available
Classic	High	Easy	Good	Based on Plan	Power from phone lines

I *cut the cord* a while back during a career transition. And I don't watch much live sports. Service provider package options are not as sweet for me because I am a data only customer. Sure, they would happily help me out, but I am content to do more work for bigger savings.

A word about power outages. If you have anything but a wireless phone or classic twisted

Companies like APC make *uninterruptable power supplies* (UPS) that store energy in a battery and can provide AC power to a device such as a computer. These are rated in Volt Amps (VA) which is like Watts but is not. They are designed to keep a computer running long enough for you to shut it down, typically two hours.

If you put a smaller load on a UPS thinking that it will last longer you will be in for a bad surprise. The efficiency of the converter drops and the time it can provide power is significantly shorter than you might think.

Using a UPS in lieu of a battery does not work. Many government regulators apparently don't understand this based on meetings I have attended.

pair, you will lose phone service when the power goes out. At the average US home, you lose power 1.1 times a year for about 1.5 hours. The power companies are not required to track outages of less than five minutes. These short outages are more frequent. These are the ones that screw up your clocks. Third party VoIP providers have no requirements for working during power outages. If your Internet goes down, and it will, you lose phone service. Your basic wireless device runs off battery to begin with and you can always buy a battery pack with USB charging interface to give you more time. Your traditional two wire phone is powered by batteries at the phone company. Fiber Optic, DSL, Cable Modem(eMTA) all are considered *wired services* by the FCC, and they must provide a device that can take a battery that last for eight hours. You will have to buy the battery.

Fax

Fax machines portray themselves as simple. But I see people struggle with them. If this is you, then read on.

People have been trying to send images over telegraph since the mid 1800s. The machine that started the modern fax system was created by Xerox in 1964. Fax or facsimile was huge in the 70s, 80s and 90s. My employer would exchange

Figure 20 Fax Machine Wikipedia.org

hundreds of pages of fax every night with our Japanese manufacturing partner in the 90's. Our VP of Engineering took a portable fax machine and a case of paper when he visited his in-laws at Christmas. Then emails came along and the popularity of fax has waned. But if you cannot scan, fax may be a good alternative.

Fax uses your phone line. If you have only one, you cannot fax and talk at the same time. You cannot fax while talking to tech support unless you have two lines.

Mark's guidelines for successful faxing:

- If you use it often, get a dedicated phone line for the fax
- If you receive faxes without warning, get a dedicated phone line for the fax.
- If you have one phone line, turn receive off on the fax machine unless you are expecting a fax in the next few minutes.
- If you are trying to send a fax, you cannot talk on the phone.
- Make sure that the fax machine you want to buy is one you know how to work. Buying a cheap fax machine that is hard to use is not saving you anything.
- Have the number you are faxing to written on a piece of paper in your hand. The cover sheet with the number on it is often face down or scanned into the machine when you need it.
- Pick a model that lets you listen for a few seconds to the terrible tones that two faxes make when they are talking. If you hear "Hello" instead of the computer-generated tones, you called the wrong

number or the recipient has their fax
turned off.
- Learn to scan and send attachments via
  email rather than fax.
  - Sorry, some doctor's offices and
    insurance companies are still in the
    dark ages and only receive faxes.
    You can fax at FedEx Office stores
    and most office supply stores.

Scanners are cheap. You put in the paper it scans
it just like a fax machine. The machine is
connected to your computer, (You must have a
computer.) and you get a PDF file, (Portable
Document Format). There will be software on
your computer. It will save the file somewhere.
Either tell it where you want it, or remember
where it saved it. Then write an email to your
destination and click on the paper clip
somewhere at the top of your email program. It
will open a file dialog box that will let you search
for and attach your PDF file. Then you hit send.
Congratulations you just moved into the 21st
Century.

The Neat Company makes a nice desktop model
scanner with nice software. I like my Brother
scanner. And Epson makes nice scanners. I have
used them all. Look at the software more than
the scanner.

I have also found a program (app) for my iPhone
and tablet called Scanner Pro. It lets me use the
camera in my phone or tablet to scan documents.

It will correct for bad lighting and straighten the document. It can even do optical character recognition (OCR) to convert the image to searchable text. It can save the document to memory or send to the cloud.

Chapter 5 Audio

Audio has moved to streaming. Even my 80-year-old father-in-law loves his Pandora.

But I still like my iPod.

Another trip through Memory Lane

When I was young, my grandparents had a small collection of 78 RPM phonograph albums. My first record albums were 33 1/3 RPM and 45 RPM (revolutions per minute). If you wanted to record sound the only consumer option was reel to reel tape. The other way we consumed music was the radio. AM was the format we started with, but eventually

Figure 21 Audio Speaker Pixabay.com

the superior sound of FM broadcasts won out. In the late 60s and 70s, alternative rock stations were the primary stations on FM radio.

For a time 8-track tapes were popular. I never liked the format and instead went with cassette tapes. Eventually music went digital and various formats came and went. The one that stuck was CD, Compact Disc.

The next game changer was the Apple iPod. Various digital music players including Microsoft's Zune player also were introduced but the iPod came with the iTunes store and the total music ecosystem. I had to look up the Zune player to ensure I remembered the name correctly.

Since then music has gone to a streaming model. Rather than own the music you borrow or rent it. It is delivered over an internet connection. The music player is now part of your cell phone.

Sources

Vinyl

Vinyl records have seen a resurgence of late. Along with tube amplifiers and classic music technologies of an earlier time. Vinyl records are analog. You fussed with them to get dust off and they eventually get scratched.

Figure 22 Record Player Pixabay.com

If you want to go back to this bygone era, there are good turntables (record players) available. You need to put the turntable on a very sturdy shelf to avoid vibrations. (Not the same shelf holding the speakers.) The turntable has a cartridge that produces a low-level signal. If your amplifier/receiver does not have an input marked *phono* (phonograph), you will need a *pre-amp*. Back in the day, all good receivers had phono inputs, but alas, no more.

CDs

The vinyl records of our youth gave way to Compact Discs (CD). These silver shiny platters are *digital*. The sound is stored in '1s' and '0s'. Each word indicates a piece of sound and if you string them together and smooth them out fast enough, it becomes analog. Because it is digital

Figure 23 Compact Disc
Pixabay.com

you can make perfect copies. The ability to make perfect copies made it easy to share and even pirate music.

Some people say they can hear the difference. But my ears are not what they were when I worked in a recording studio. CDs are good enough for me. There are other less common formats that use more '1s' and '0s' in each word and they use more words (samples) per second of audio. Some even have 5.1 channels of audio. But if you know this you are beyond this book.

A word of caution: One side of the CD has a label. The other side is clear and you can see the tracks. The CD player shines a laser through the clear plastic to read the tracks. But the data is on the other side of the disk under the label. Scratches on the label side can destroy the data.

I built a big cabinet to hold all my CDs. I acquired a 400-disc CD player. And still I had CDs scattered about with nowhere to store them. Before I switched to downloads, I started *ripping* (converting the .wav format to a different format) the CDs into MP3 and other format files and storing them on my music player (iPod) The plastic cases I threw away. I bought a suitcase designed to hold hundreds of CDs. Like scanning photos, it is an ongoing process. I don't know that I will ever convert them all.

Downloads

I've cut out the middle man now. I simply go to my favorite music site (Apple, Amazon and other smaller sites) and download my new music. This works for old music too. I tried converting my mother's Glenn Miller vinyl albums to digital and the quality was

If you want to convert your old vinyl albums, you can get a turntable (record player) with USB output that will connect directly to your PC. There are programs like Sound Soap that can process those files and clean up some of the pops and clicks. But it is time consuming and requires some tweaking.

awful. But I was able, for a reasonable price, to purchase digital versions from iTunes and later convert those to CD format (.wav) for my mother who does not have an iPod. I say this but my 91-year-old mother has expressed interest in an iPod just this last week.

Satellite

There are 2 broadcast music services today. One is available from your cable TV or Satellite TV provider called Music Choice. Music Choice is available with nearly all your digital video subscriptions. If it is not in the basic package it will almost certainly be in the

Figure 24 Satellite
Wikipedia.org

next better package. It will have 30 or more channels with music for every taste. If you connect your cable or satellite box sound output to your stereo system, you will get better sound than using your TV.

The second satellite service is Sirius/XM. This used to be two separate services but they combined. They have hundreds of channels including most major sports, news and comedy. Most cars above a certain price point come with Sirius/XM radios built in. Some Stereo and home theater receivers have Sirius/XM radios built in. And various standalone radios are available. The service is nice. It costs between $11 and $20 per month depending on how many channels you desire. Additional radios are extra. It will tell you what song is playing and who is performing.

Internet Streaming

Many people are *streaming* music today. Spotify and Pandora are the granddaddies of them all. Apple has a streaming service and the list goes on. With Spotify, you tell them one or two artists or songs and you

> In babysitting my new grandson, I was gratified to find that Spotify has a "Raffi" children's channel.

get a playlist of similar music. Your typical streaming service is free with ads. They have a monthly fee for no ads. All songs are not available on every service. You should shop around.

The down side to streaming audio is that you must have an internet connection. No Internet means no music. If you are streaming on your phone, it is using data minutes and your bill could be large depending on your contract. This can go to ridiculous extremes if you are traveling internationally and roaming. Despite the potential for extra cost and complexity, my father-in-law loves his. I'm sticking with my satellite and iPod.

Digital Music Formats:

There are 2 basic types of digital music formats: "Lossy" and "Lossless". Lossy file formats typically have smaller size by throwing information away. Before you recoil in horror, it tries to do this intelligently by only throwing away the stuff that your ear and your brain tended to ignore anyway. If you optimize for smaller files sizes, you can lose noticeable audio quality. Lossless compression does not throw away any audio data but uses file compression technology to remove redundancy and achieve smaller files. Lossless files are larger than Lossy files. Small file size was important back when music players had very little memory. Today, memory is cheap. I recommend Lossless files.

MP3: (MPEG1/2 Layer III.) Is a popular lossy format. If you stick with the higher bit rates like 192k it sounds OK, especially in a noisy environment like a car or airplane.

AAC: A newer lossy format. It is less common.

Apple Lossless Compression: This is the default iTunes format and is lossless. Even though no audio data is thrown away, the files are about half the size of uncompressed files.

FLAC: This is a less common format. It is lossless. FLAC stands for Free Lossless Audio Compression.

WAV: This is the format of the digital music on your CD. It is lossless.

AIFF: A less common lossless format.

Equipment

Back in my youth, when you moved into a new dorm room or apartment, you set up the stereo first. Your bed, clothes, pretty much everything else was subordinate to the phonograph, amplifier and those huge speakers.

Times have changed.

Stereo anyone?

Two-channel sound has become less popular in the home environment. If your sound system is part of your TV experience, it makes sense to use a home theater configuration. These home theater systems are designed for 5.1 or more speakers. But two-channel systems are still available.

I'll bet that most of your music is still two-channel (stereo). CDs, iTunes and Amazon downloads, streaming media, are all two-channel. You can get some music in 5.1. I tried DVD-Audio when it was available in my Acura automobile. It sounds great but the number of titles was limited and frankly was more trouble than it was worth. But be aware that it is an option.

Most headphones are two-channel stereo. If you want to listen without forcing others in the room to listen to the same sounds, headphones are a good option. There have been attempts at making them 5.1. In my opinion, there is too little room even in large headphones to do a good job with 5.1.

In buying that two-channel system, good speakers will cost more than anything else. The classic recommendation is budget half your money for speakers. Go to a good audio store and listen to various options. You must move air to make sound. And bass notes have a longer wavelength that typically benefit from a larger speaker. In my house, big speakers are banished to the home theater. But there are ways around this limitation. You can get nice in-wall or in-ceiling speakers that can be painted. And to improve that base, think of a dedicated sub-woofer. Bass is non-directional and is typically not stereo. One subwoofer can really help that bottom end. As in the dorm room or apartment, running the wires will be your challenge.

Your local stereo store cannot really compete with the big box stores on price. But I hope you will give them business, because where else can you listen before you buy. Also, they have installation services. If you do buy on the Internet or a big box store, you can still pay the local stereo store to help you install your gear.

You might be better off with a wireless solution like Sonos.

Before you buy, think about where your music comes from: CD, iPod, Internet, Cable or Satellite set top box, Bluetooth? Do you still have a turntable and 33 1/3 RPM albums? Do you want Sirius/XM? I swapped out the receiver in my living room just to get a different selection of inputs. I have Sirius/XM but I use an external receiver for that. The Sirius/XM satellite antenna needs line of sight to the southern sky. But you can also get the Sirius/XM app for your tablet or phone and connect to the receiver with Bluetooth. Modern receivers can connect to the Internet. If you don't have an Ethernet jack behind your setup you can get a Wi-Fi *dongle* to plug in to most USB ports. Units with internet connectivity often come with streaming audio services.

Some of these features require a display to set them up. They might assume a TV display will be connected. But unless this is part of a Home Theater setup, there may be no TV. You can get a small TV with HDMI input for about $100 to use short term just to get it set up if it is not normally connected to a TV. If you do this, make sure the TV also takes a computer input. Then you have a second monitor to use with your computer when you are not setting up your receiver. There will be a wide variety of other two channel inputs with various names. Some units let you rename them. I rarely find this useful.

 Be careful if you intend to use a turntable. The *phono input* is disappearing. Your turntable puts out a much weaker signal than other audio devices. It needs a boost. Older receivers have inputs labeled phono. These inputs boosted the signal from the turntable, aka phonograph. Never plug other audio devices into an input designed for phono. If you purchased a receiver without phono inputs, all is not lost. You need to purchase a dedicated pre-amp to boost the signal from your turntable and feed it into any of the normal inputs of your unit.

The pre-amp, whether it is built-in or external, does something else. The frequency response of the turntable is very different from a CD player. Frequency response is the balance between high notes and low notes. The pre-amp changes that balance to make the sound *normal* or more pleasing. This is another reason not to try plugging your CD or other audio source into inputs labeled phono. Plugging a CD or other audio source into a phono input would make it sound strange and it would be too loud.

Bluetooth

Bluetooth is a two-way radio technology that is very short ranged, 30 feet. It is the radio used to connect your smart phone to your car and wireless headsets. The name is cool. It is named after Harold Bluetooth, the Danish King, who united Denmark in the tenth century. It is low power and ideal for some audio applications like connecting your tablet to your stereo receiver. You can then have SiriusXM or

Figure 25 Bluetooth Logo Wikipedia.org

Spotify or other streaming media on your receiver with little hassle.

Warning, if you connect it to your smart phone to play music, your phone calls will end up on the audio system. This may be OK if you are by yourself, but it will interrupt a dinner party in a heartbeat.

Home Theater

When you pay your $15+ to go to a theater, you experience the big screen and immersive sound. Today, bigger TVs are giving you that big screen. But the other half of the experience is the immersive sound. This is where 5.1, 6.1 and 7.1 sound comes in. The Japanese and some movie theaters are working on a 22-channel sound technology. But I don't expect home owners to buy that many speakers or run that many wires.

Legend has it that the rear speakers were added so that the Space Ships in Star Wars would sound realistic when flying from behind.

The '.1' refers to the subwoofer for low notes. You still have left and right front channels. That gives us two. There is a *center* channel used for dialog that doesn't feel right coming from right or left. We are up to three now. There are left and right back channels. These are typically placed ear level to the sides of your prime listening area such as a sofa. Now we are up to five. With the Subwoofer that give us 5.1. Because the rear speakers are good for ambiance and because not everyone can sit in the center of the couch at the same time, some systems add a true rear channel directly behind the listening area. This rear channel can have one or two speakers. Listen to 'Gladiator' in 7.1 and you hear the crowd roar go around the colosseum like the wave at a baseball game.

Big speakers are great. They move a lot of air. I like a bright clear sound for movies and choose a different speaker for my home theater than for pure music. Many homeowners don't want to accommodate the big speakers anymore. There are systems with small speakers like Bose that use electronics to give a pretty good sound with smaller speakers. And don't forget, you can mount some speakers in the ceiling and walls.

Dolby Atmos

This is too new and too complex for this book. But I can tell you briefly what it is. Dolby has created a way to record sound that can be played on different systems with different number of speakers. The home version of it needs speakers in the ceiling or speakers that bounce the sound off the ceiling. This adds the vertical dimension to your surround sound. You can get by with two speakers for the ceiling but Dolby recommends four.

And with more speakers you need more channels in your amplifier. If you are moving to UHD, you needed to upgrade your home theater amplifier to handle HDMI 2.0 and HDCP 2.2. Some systems allow the extra channels to be in an external amplifier. You will see configurations like: 5.2.4 or 7.1.2. The first digit is the number of traditional surround sound speakers. The middle digit is your subwoofer(s). The last digit is your ceiling speakers. And the amplifiers decoder must understand Atmos.

TV Sound

When I worked in a TV studio, our sound mixing room had nice expensive *studio* speakers. It was a pleasure to use. However, when the director would come in to review the sound, he would hit the switch and turn off the big speakers and run all the sound through a little 6-inch speaker. Because "That's what it sounds like at home from your TV." And today, things have gotten better and worse.

Thinner TVs don't have room for that six-inch speaker and don't have big volume you can harness to boost the bass. Truth be told the engineering in those TVs is amazing, that it sounds as good as it does. But it doesn't really sound all that good. Solution, get a sound bar or connect the TV to your Home Theater setup.

The improvement comes from sound bars and Home Theater setups.

Whole Home

You have two choices in whole home, wired and wireless. If you are building a house it can pay to think of putting in wires for audio. I recommend putting in 18 or 16-gauge lamp cord wire in the walls. A local sound advisor can help you plan the installation.

For most people with an existing home or apartment, running wires is not the easy option. For years, the holy grail of home speakers has been a wireless speaker, but for one reason or another a solid solution has been elusive. Manufactures have finally solved the problem, beginning with Sonos and now from Bose and other companies. These speakers use Bluetooth to connect to a smart phone, iPod, or tablet. Some connect directly to streaming apps like Spotify, Pandora, iTunes or Amazon. The small speakers are about $200 each, the medium about $350 and the large speakers are about $500 each. Double that if you want 2 speakers for stereo. They talk to each other using your home's Wi-Fi. If you don't have Wi-Fi then you need wires for a whole home setup.

There are a wide variety of small Bluetooth speakers that allow you to listen to music from your smart phone or tablet. Most have the option to plug in an iPod music player. But these are not considered *whole home*. Some are good and some will pair with a second speaker for true stereo. I use a little Jabra unit when I'm traveling. Amazon Basics has one for about $20.

Portable

I have an old classic iPod with 160GB disk drive. Most people use their iPhone as a player but you can still get a new iPod Touch with anything from 16G to 128GB. And they make smaller players now. But my old classic holds all my recorded music and many of my audio books at the same time.

Figure 26 iPods
Wikipedia.org

Chapter 6 Photography

The *best* camera is the one you have with you.

This is likely your smart phone.

Figure 27 Kodak Instamatic Camera
Wikipedia.org

This is probably one of the biggest changes we have faced. My first camera was a Brownie. It used 120 (or was it 127?) roll film. It produced little square pictures. In the early 60's I graduated to the Kodak Instamatic. I think I had a Kodak Instamatic 100. With the little peanut flash bulbs. It used 126 film cartridges. Mom had an Agfa 35mm range-finder. She shot Kodachrome slide film. We knew when dad took the picture because the top part of your head was cut off. My little sister got one of the little 110 instamatics that were all the rage. But not me. The bigger the film the better, right?

Who knew that Kodak would lose the market and be displaced by a phone. Pictures? Phones? What do they have in common. Times have changed.

What is the best camera?

I wrote a photography blog for a couple of months. It required more writing than I had time for but one of the questions that people submitted to me was "What is the best camera?" Usually followed by their opinion of Canon, Nikon, Sony, and others. My simple answer was the best camera is the one you have with you. Life happens fast. If you don't have your camera with you, you cannot take a picture of it. If you always carry your pro SLR camera where ever you go, that is hard to beat. If you run out for pizza and you see an amazing sunset or someone wrecks your car, that camera on your cell phone starts looking pretty good.

When we went on vacation, before cameras on phones got as good as they are today, I would carry a small pocket size camera in my pocket so it was always with me. When we went to see the sights, I had the big SLR (single-lens reflex) digital camera.

Early on, the Nikon vs. Canon debates raged among the *discerning photographers* who used SLRs. The problem is, unless you are rich, when you buy your first SLR you are wedded to the lens system. And your investment in lenses (glass) overwhelms the cost of the camera. Once you are in one camp, leaving it becomes cost prohibitive. So, like the song says, "Love the one you're with." They are all excellent or they would not still be in business.

Choices

We are going to talk about digital cameras. Film is dead. Unlike vinyl records, I foresee no resurgence for film. I see articles about guys who buy out lots of film when they find it, and store it in a freezer for use later. They want to keep that *film feel*. This is not average behavior. We are not addressing them. They know what they want. It's crazy but, hey, whatever floats your boat.

This is where we ask what do you want to do? Rather, how will you view your photograph? On

> *Before we go forth we need to discuss Mega Pixels. Mega is shorthand for 'million'. A 'pixel' is a picture element, a dot. The way people talk about them you would think more is better, higher resolution and all that. I'm asking you to trust me on this, many factors go into the quality of a digital photo. MP is not first and foremost. Start by ignoring MP numbers. I'll do a fuller explanation in the back of this chapter.*

Facebook or Instagram? In that digital picture frame? Family web site? Prints? How big a print? If I'm at a concert, or kayaking, or hiking and I want to use Facebook or Instagram then taking a photo with my phone is good enough and it gets stored on the cloud somewhere. I can send it straight to Facebook from the phone. If you want to strap a camera to yourself, your bike, your surfboard and take movies of great action shots then a GoPro mini-video camera is a good candidate. If you want to climb a mountain, find just the right spot and capture a sunset and blow it up to humongous proportions and frame it. Then you at least need a Single Lens Reflex (SLR)

Lenses

We need to discuss terminology before we go forth. In Chapter 1, we became familiar with the power consumed by incandescent light bulbs. 40W was dim. 60W was OK. 100 W was good for

reading. But that said nothing about the light they put out. It was the power they used. When you switch technology, the metrics people are comfortable with change. And you start to see words like "equivalent". This bulb is *equivalent* to a 60W

Figure 28 Camera Lens Wikipedia.org

bulb, meaning it put out 800 lumens of light. Camera lenses are the same way. People got accustom to 35 mm film cameras. That is the *standard* for how photographs look with a given lens. Incandescent is the *standard* for how much light is produced.

If the ad says, the equivalent lens is 30mm to 120mm. Rest assured the lens on a phone or compact camera is not 30mm to 120mm. When someone says, "The camera has a zoom lens that is equivalent to 35 to 140mm..." They mean that the pictures it takes will have the same perspective as a photo taken with a 35mm film camera with a zoom lens that goes from 35 to 120mm.

In the land of 35mm film cameras, a 35mm lens or less is wide angle. Meaning wider, taking in more scene, than normal human vision. A 50mm lens is normal. A 100mm or greater lens is telephoto. It brings far things close. When you see *equivalent*, the gold standard is 35mm film.

The amount of light that a lens brings in is controlled by the aperture. Basically, the size of the hole. The smaller the aperture number, the bigger the hole and the more light. (think golf, low score wins) A lens with a 1.2 aperture is fantastic, expensive and heavy. A lens with 2.8 aperture is more consumer grade and lighter.

A personal word of advice. Taking great photos can be a nice hobby. But don't let it get in the way of relationships. Some vacations are better used relating to our families and not chasing the next best sunset.

Phone

In theory, this camera is ALWAYS with you. It usually has a camera on both sides to make selfies easy. It is connected to the Internet. You can send a photo to Facebook now, instant gratification. It will take movies, normal, square, and pano (long photos you get by moving the phone.) And with simple software you can get all sorts of effects. Did I mention, it is always with you? **Caution:** sometimes an 8MP camera can take a better photo than a 12MP camera.

Sensor size on a phone is small. So very small they don't even quote the numbers. Photography means "writing with light". The larger the sensor, the more light you capture. Big sensors are good. (and expensive) What phones lack in sensors they make up for with amazing software. And did I mention, you are connected to the Internet.

Compact digital cameras

On a compact digital camera, the sensor size is larger. Larger sensors capture more light. Because this is a dedicated camera, it will have a bigger and better lens. It will often have mechanical zoom. (*Digital zoom is simply throwing away pixels.*

Figure 29 Digital Camera Wikipedia.org

Ignore it. Don't use it. Crop later.) With a larger sensor size and a bigger lens, it captures more light. My latest small digital camera is water proof. It takes great pictures. It can get wet. It has zoom but can fit in my pocket. All goodness in a compact digital camera.

An advantage to a dedicated compact digital camera is that you can get waterproof, dust proof and a wider mechanical zoom than with your phone. It's one more thing you must carry, but maybe you don't want your phone in the water. Some cameras have GPS to record where you took the photo. But GPS will eat batteries. Again, there is no free lunch.

SLR

This is the direct descendent of my mom's Agfa Range-Finder. It mimics a 35-mm film camera.

Figure 30 SLR Camera
Wikipedia.org

SLR means Single Lens Reflex. It has a complicated mirror system that lets you look through the lens and see the same thing the sensor will see. (Mom's range finder had two lenses. You looked through one and took the picture through the other. If you did not compensate it was easy to cut off the tops of people's heads. You are forgiven dad.) Some cameras now have digital view finders. You see the image taken from the sensor. But the resolution of the display does not match the sensor. For snap shots, no problem. If you get advanced and want to do arty things, then this might be a problem. *[When you reach this stage, you are beyond this book.]*

The key thing about SLRs is that the lenses are interchangeable. You can switch from wide angle to telephoto to more exotic lenses such as fish eye or macro. Trust me the variations are only limited by your wallet. The lenses are specified as if this was a 35mm film camera. I should also mention: the sensors are larger than the sensors in compact digital cameras.

Some SLRs have smaller sensors than 35mm. A 50mm lens looks more like a 75mm. 35mm is closer to normal than wide angle. The more expensive SLRs have 35mm sensors. These are called *full frame*. The bigger the sensor, the more light it captures. The more pixels, the less light per pixel. The less light per pixel, the harder it is for low light and the more noise you are likely to encounter. More pixels are not without problems.

Once I was working with Kodak and I had a VP tell me that cameras would never have more than five Mega Pixels. His reasoning was this tradeoff of pixel size and the ability to capture light. In fairness, he was talking about small consumer cameras.

SLRs have other features as well. Most SLRs today can take HD video. You have a great deal of artistic control in the tradeoffs of aperture vs. shutter speed, depth of field and stop motion. Or you can create a sense of motion with a slow shutter speed. You move from event capturer to artist. And that is why people spend tons of money on big cameras.

Go Pro

Among the younger typically more athletic set the perfect camera to capture their feats of derring do is a movie camera with a wide-angle lens. Several exist but the GoPro has set the standard. There are different models and the top of the line is the GoPro Black that shoots 4k Video. It comes with various gizmos so that you can mount it on your surfboard, bike,

I have a friend that did not secure her Go Pro to her wrist or something that floated. It ended up on the bottom of a very deep lake. Be careful!

helmet, chest, or drone and get first person videos or selfies. It has an optional water proof housing that makes it good for kayaks, surfboards or snorkeling.

The lens is very wide angle. Great for action. But if you are trying to capture people you need to get close, very close. If you want to capture sound like the howler monkeys in Belize, don't use the water proof housing. If you want to shoot underwater, invest in a red filter. *[Personal experience on both counts.]*

It comes with good software. Your biggest problem will be finding music that YouTube does not recognize as copyrighted material for the sound track.

While some drones (hobby helicopters with three or more propellers) can take a GoPro, more expensive models have built in cameras that are less wide angle.

Post processing

Every photo can benefit from making a few adjustments. Correct the lighting, crop the picture to get close when the photographer forgot or could not. Get rid of *red-eye*. Add an effect. Give the blues a bit more punch. If you used film back when, and took it to the drug store for processing, you didn't have this option. You were only able to adjust your photos if you had a dark room. Now, everyone with a digital camera and a computer can play.

The biggest mistake that people make is not getting close enough. The best zoom lens is often your feet. If you can't get close, then use your software to crop out things that are not your subject. Notice the ads for cameras always have close people shots.

There are programs on your phone that you can use to fix and spice up your photos. Try your photo app. And then go to the app store and go shopping. Search for photo. Your tablet is bigger, easier to see but has the same software as your phone if they are from the same manufacturer.

On your PC, both Apple and Windows have native photo programs. I use Apple Photo for run of the mill, quick and dirty adjustments. If I get serious, I use Adobe Lightroom. It is an amazing program designed from the ground up to be for photographs. Photos and Lightroom are each a database and adjustment tool. Lightroom is not too expensive at about $142 or $9/month for the subscription option.

If you want to really go crazy and spend an insane amount of money, there is the venerable Adobe Photoshop. It is the ultimate in pixel manipulation. Added to Lightroom, there is no limit to your creativity. Well, technically, *YOU* are the limit. With power to do virtually anything comes a very intense learning curve. I should mention that Adobe has changed its business model. They greatly prefer to rent you the software for monthly or annual fee rather than simply selling it to you.

Recently I found Pixelmator for the Mac. Pixelmator does about 90% of what I did with Photoshop. It cost about $30 US.

Viewing

In 1975, I spent the summer in Europe with my cameras and 300 feet of Kodak Ektachrome color slide film. (The same film used by National Geographic for natural light shots.) I came home with 2,000 usable images. I dutifully mounted these in plastic mounts and then sorted them into the 200+ images I used for a presentation. The other 1800 images have only been seen by fewer than a dozen people. My point is: who and how will people see your photos?

Social Media

My wife and I joined Facebook (FB) to spy on... I mean, keep up with our kids and now grandkids. All our neighbors joined FB, and then even our parents. And FB is the way we share daily moments. Selfies at a rock concert. Celebrating an anniversary at the local Italian Restaurant. Plugging a favorite hotel or a friend's beer. The target of 90% of our photos is social media. For that, my phone is more than good.

My daughter has dragged us kicking and screaming into yet another social media app, Snapchat. She will put a photo or movie of our grandson on about once a day. Of course, we will not miss that. Snapchat deletes the photo after viewing. My daughter is fine with this. Not all photos or videos are of immortal quality. My wife hates it. She would keep them all if she could. And Snapchat allows you to add cartoon features and other artsy stuff to your photo or video.

Prints

When we come back from vacation. My wife wants 25 4x6 photos (just like the drug store of old) to take to work and show her buddies, "This is why I was gone for a week". A phone or compact digital camera is fine for taking these photos.

Figure 31 Prints
Pixabay.com

Some vacations I don't take more than my phone, compact digital or GoPro video camera. I choose to focus on the people I'm with. You can do the same; you have my permission. This is OK. But for that life-time trip to Hawaii, I take my SLR. I get excellent photos and blow them up to obscenely large size and hang them on my wall. (Because I can. I'm embarrassed to admit, I have a photo printer that can print up to 24" by

99". The frames cost more than the prints, so I build my own frames.)

Books and Stuff

When we take that 25th anniversary vacation or go to an exotic locale, regardless of what instrument took the photos, I like to put the best photos in a book. The easiest options are not the cheapest. Apple will let you create a book from within Photos (formally iPhoto). Shutterfly, Mixbook, Pinhole Press and Snapfish are popular online services. My favorite option is Blurb. www.blurb.com . Search for photo book on line and see what the different options are. The easy or popular services can cost twice as much as a company like Blurb. It pays to shop around.

Our friend, Kim, takes the time to create a photo book each year for each of her children and their kids. It is a wonderful record of her family's growth. Now that I'm a grandparent, I need to find the time to emulate her stellar example. Having thousands of random photos without being edited and curated into a more concise form is a guarantee that no one will ever look at them.

Electronic

Some TVs have a USB connector that can be used to plug in a USB memory stick. If that stick contains photos you can instruct the TV to play them. You might need to have only JPEG photos in a single directory. *[See later paragraph on digital formats.]* Be careful here. Even with close family and friends, less is more. Don't try to show all 200 just because you can. You can thank me later.

Another way to show photos on your TV is to use a program like Apple photo to play a *slide show* but you will need to connect your apple device to the TV. You can use an HDMI interface as an optional accessory or use *Apple TV* and *Air-Play*. Apple TV can also show photos in your Apple *iCloud*.

Maybe you think that a TV is too big, too limited,

too power hungry to be used to show your family photos. You just need a simple frame. You can purchase a *Digital Picture Frame*. Prices range from $40 to $150 for a reasonable 8" to 15"

Figure 32 Electronic Frame Wikipedia.org

picture frame. You plug in a memory card that you loaded with photos from your computer or camera and they will display. You should have a choice of one photo per day or a slide show with 5-30 second intervals. Some frames can also show a clock or calendar. Some frames show movies. Be careful that the frames software can display the format you want to display. JPEG should always work.

Storage

I have boxes of color slides from my mother capturing my early years. I have boxes of old Black and White photos from the family. I have boxes of color photos of my kids growing up. The old family photos are fading. My mother's slides are fading. There is no organization of my family photos. The wedding photos in our fancy wedding book are starting to fade even though they are not exposed to light. Traditional photographs and even slide transparencies will degrade with age. Digital photographs have the potential of remaining timeless, unchanged. But you need a plan to store them.

Converting photos to digital

The good news is that all my photographs for the past decade have been digital. But I have a century or more of family photos that I need to move to a digital form. This is a straight tradeoff between time and money. If you have more money than time, there are services that you can pay to convert your negatives, photos and slides. Search on line for 'convert photos to digital', and you will find a wide range of options. Adding the word 'review' to the search will get you reviews from sources like CNET, PC Magazine, USA Today and other sources. Pay attention to the date of the article. You often see reviews that are several years old and likely out of date. *[This is true for most online searches.]* PC Magazine had a nice write up. Expect about $0.50 per image. Prices will vary based on your choices. If you only do a few at a time, then shipping charges will drive the price per image up.

If you have time and some minor computer skills, then the DIY option may sound better. For photos, you will need a scanner. For photos, a flatbed scanner works well. Some offer attachments for slides. But I have seen better results on line using a slide converter with a built in digital camera or using a digital camera you already own. The flatbed scanner should be about $80. The dedicated slide scanners seem to run in the $100 to $200 range with the highest rated in 2015 costing around $120. You will need a computer for this work.

Do it yourself or pay a service, either way, let me make a few suggestions. Not every photo you took is worth the effort to convert it. As I was working with Mom's slides, many are of places they visited. None are worth framing. Time has taken its toll. But there are pictures of Mom and Dad. Pictures of my sister and me as kids, family Christmases. These are worth preserving.

- Step #1 is to edit. Throw out or at least set aside the photos that are not worthy. Be brutal in editing. You will save lots of money and time.
- Step #2 Clean them. Use compressed air and a soft cloth that you might use on glasses.

- Step #3 Organize. Preserve the context. Country, date, event.

If you send them to a service. Start with a few and make sure you are happy with them. Split up different categories. What if the package gets lost? I assume that if these are worth the cost of conversion that these are precious memories. Minimize the damage a single service or package delivery can do.

Storing Digital photos

Digital photos are great. They do not age. You can easily edit them without a darkroom. But if the disk or device they are stored on goes bad, you lose thousands at a time. What if your home burns down? The good news is that you can make copies without losing any quality. You can share those photos with family. You can store them in multiple places on multiple devices. But you need a system.

Organizing

I have a directory on my computer called Photos. I did not create it. Apple provided it. But it is a very good place to keep all my photos. I have directories that I dump photos into so that they are organized. 20 years of family photos are by kids and years. But those are duplicates. I also have photos in an image organizing software. I use two, Apple Photo and Adobe Lightroom.

Like anything else, there is a lot of image organizing software available. If you use apple products, you already have Photo. Adobe Lightroom is more advanced. Google has a good free program called Picasa. I'm sure Microsoft has something built into Windows. Image Organizers allow you to find, adjust and organize your photos. Many have facial recognition and can sort photos by the people in them.

Storage Media

When digital photography for consumers began, file sizes were small. Cameras had few megapixels and stored images in compressed JPEG format. It was easy to store your photos. Today, with lots of pixels and uncompressed raw formats it takes a large storage device to hold your photos.

There is no perfect choice. All choices have tradeoffs. The good news is, while photos are bigger, so are disk drives.

DVD-R

I skipped right over CD-R. At 700 Mega-Bytes (MB) it does not hold enough of today's photos. A DVD-R (recordable) holds 4,700 MB. My guess is that it will take several of these to hold your photos. The problem is that they can be damaged, and formats go out of style. Many computers today no longer have a CD or DVD reader. Good news is that a DVD reader that plugs into your USB port is cheap. I backed up my iPhoto to DVD a few years ago and gave copies to Grandma. If my house burns down, she has the family photos.

Hard Drive

This is the easy option. My guess is the most common interface (wire used to connect to the computer), USB will be around for a long time. You can get an external hard drive with 2TB (terra bytes or 1,000 Giga bytes or 1 million Megabytes) for about $80. Just copy your entire Photo directory. Remember to do this every so often and you can keep it in a fire proof safe or safety deposit box. Better yet. Get 2 and alternate.

USB Memory Stick

I feel better about USB memory sticks in fireproof safes. I have no data to back this up. But you can get 512 GB for about $300 today. Think of this as a solid state hard drive. They are small, about the size

Figure 33 USB Memory Stick
Wikipedia.org

of your thumb (aka thumb drive). A word of caution: Get a name brand like Lexar, SanDisk, Kingston and not a bargain brand. I was using a bargain brand to transfer files and have seen failures after several weeks.

Cloud

I said in the chapter on Internet that the *Cloud* was simply someone else's computer and storage that you could use over the Internet. Apple, Google and others will give you some storage for free when you set up an account. Microsoft Office 365 business comes with a Terabyte. Companies like Dropbox will do the same. And there are companies that cater to the *Cloud Storage* market. If you need more than the free space, these companies will sell you more space for a nominal fee. Some of the original pioneers in Cloud Storage went bankrupt and went away along with your files(photos). I'm confident that large companies like Apple, Google and Dropbox will be around a long time.

The Gory Details

OK, this is not average person stuff. It can be very technical. If you are interested, read on.

Digital Image Formats

The sensor in the camera does not actually record color. Each pixel records light, all light. To

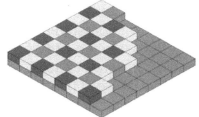

get color, the camera uses a bayer filter in front of the sensor. Each pixel has a filter that allows that sensor to measure one of three colors, red, green or blue. The bayer filter is named after Bruce

Figure 34 Bayer Pattern
Wikipedia.org

Bayer a Kodak employee, who filed the patent in 1976. *[There are twice as many green pixels as red or blue to mimic the behavior of the human eye.]* Each pixel measures brightness, and the color is computed later by looking at the surrounding pixels. *[The human eye is more sensitive to light intensity than color.]*

The most common format is JPEG (Joint Picture Expert Group). It is a compressed format that stores eight-bit RGB information for each pixel. Other compression formats like PNG are available but JPEG is king. JPEG is a Lossy format. It throws information away to get a smaller size. Depending on how it is viewed, and how hard you compressed it, you may not notice the missing information.

The information coming from the camera sensor has much more information than is saved in the

JPEG. Because information is thrown away, you lose some ability to adjust the photo. If you keep the full camera data, the *raw* file, then adjustment opportunities expand. Each camera has a different raw format. Computer Software must continually be upgraded to understand each. You run the risk of a format going *out of style* and not being readable. One way to solve this is to use a lossless file format like TIFF. But these files are huge. A better alternative is Adobe DMG. When importing photos Adobe converts the camera native raw format to the Adobe common raw format. There are no guarantees but Adobe is likely to be around directly or with third party support for many decades.

Chapter 7 Internet of Things

> The bottom line of Internet of Things is "Caveat Emptor", let the buyer beware.

I'm really struggling how to talk about Internet of Things (IoT) with the average person. I get asked, "What is the Internet of Things?" My friend, Kinney, said it was "stuff watching you." While not the whole story, this has an element of truth. Nest, a Google company has a wickedly good internet connected thermostat. Another friend, Tom, bought one but his wife would not let him install it regardless of the potential savings and convenience. Her view was that she did not want Google to know that much about them. More on that later.

Figure 35 Nest Thermostat Wikipedia.org

IoT is new and evolving. Read the Crystal Ball chapter to put it in a larger context. Given that it is new and evolving, I'd like to take a minute and put it in context of the technology changes that are making it possible. You can skip this intro and dive right in to the later sections, but to make it more understandable I think you need some background. I'll do my best to keep to our contract and make it something my wife and mother can understand.

First there were **things**. I'll give other examples later but I'm going to start with a thermostat. The thermostat we had when I was a kid, and for a long time after that, closed a switch when it got to a certain temperature. And you could twist the dial and adjust that temperature. In the winter, my dad would turn it down at night and back up again in the morning.

Then we had **smart things**. Engineers could put computers on a single chip that cost less than a dollar. Those cheap little computers made it possible to do tasks that *dumb things* could not do. Thermostats became programmable. I could tell my thermostat to lower the temperature at night in the winter and warm the house back up in the morning. And I could choose what time, and have a different schedule every day of the week. Not only can a thermostat be programmable but the next step was to make it smart. My Nest senses when I walk by. If it doesn't *sense me* for a few hours, it assumes I'm out of the house and goes into energy saving mode. It also pays attention to when I adjust it and will program itself. Some smart thermostats on large buildings had external weather sensors and can manage energy use based on what is happening outside. If electricity costs more during some parts of the day, it can adjust to heat or cool during the cheap times and use less power during peak times.

The next step in our evolution is smart **connected** things. Today, we are surrounded by *universal connectivity*. In our homes these are mostly devices communication using radio waves. We have 3G/4G/LTE phones. We have Wi-Fi and Internet. Our phones and home entertainment systems use Bluetooth. And there are other radios in use. *[A radio is a wireless communication medium.]* We are surrounded by smart things talking to each other. My electric toothbrush even talks to my phone and could talk through my phone to the Internet. In my thermostat example, my Nest Thermostat now uses Wi-Fi to talk to the Internet. I have a web site on my tablet device that makes it easier to set up and change the schedule by a factor of a about a thousand. I have multiple HVAC systems (Heating and Air Conditioning) and they each have a Nest thermostat. They can now coordinate their activities. And I no longer need an outdoor weather station. The Nest will get that from the weather bureau.

Connectivity has

1) Improved how I talk to the device.
2) Allows me to control the device remotely. Nice when I want the vacation cabin warm before I get there on a cold winter night.
3) Allowed multiple smart devices to work together.

4) Replaced an expensive sensor package with information from the Internet. (outside weather)
5) Allows changing the behavior of the device by downloading new, hopefully better software. (The computer program that controls its behavior.)

Any one of these things could be reason enough to make a *smart connected device* better than the two previous generations.

Warning: Sometimes IoT is a solution looking for a problem. I saw a smart washing machine that would email you if there was a problem. I don't want to be sitting in a movie and receive an email that my washing machine is out of balance. You should realize that not all smart connected things are better. All appliances have gone smart. The little computers are so much better at running things over mechanical controls. But not all appliances have figured out how adding connectivity make them better. As I said, the technology is evolving.

Smart connected devices can absolutely be much better, like my thermostat. But not all companies are equipped to figure out how to add value. And even when they do, the skills required to build the thing are very different. Expect old players to fade away and new players to step up with some cool products.

Because these are connected to the Internet, if security is not designed in and done well, IoT devices are another path to have our information *hacked* or stolen. Even if we can keep the bad guys out, Nest (aka Google) now knows many things about my life such as when I am home and what my patterns are. I gain control and cost savings while trading off my privacy. Is it worth it?

Radios used in IoT

What comes to mind when I say "radio"? Often, we think of listening to music or talk over AM or FM radio devices. When we talk about radios here, we are talking more generically about wireless communication. Examples are Wi-Fi, Cell phones, Bluetooth.

When we talked about the Internet we used a term, the cloud. We said this was someone else's computer connected over the Internet. The Internet is heavy weight communications. It is good for lots of data moving very fast. Out on the edge of the cloud moving lots of data fast is not the priority. You need small chunks of data, low power and short range. The edge of the cloud is very different indeed. We call it *the fog.*

The fog is the edge of the cloud. Things are slower out here. Low power and long battery life is king. The Fog is made up of radios. Often many kinds of radios speaking in different languages. But it is helpful to know about some of these *tribes* before we talk about how they are used.

Proprietary

Sometimes you don't want to talk to devices made by other companies. Either to shut them out or because you want something different not offered by a standard radio. AT&T uses a proprietary radio for their security sensors with Digital Life. This radio is very secure. It offers robustness against interference. And it has good security. All things you want in a security system.

Wi-Fi

Wi-Fi is nearly everywhere in every home. If you need to connect to the Internet, Wi-Fi is an easy choice. Except, drum roll please... it is not energy efficient, it is not simple and it not the cheapest radio in town. It is good for lots of data fast but most applications, even those using Wi-Fi for IoT don't need lots of data fast. But it is everywhere. It needs more computer power to pull it off. If you need a cheap sensor that will run for ten years off a coin-cell battery, pick a different radio. If you want to talk to a radio already in most homes, then Wi-Fi is a good choice.

Most thermostats use Wi-Fi. Many light bulbs use Wi-Fi. This is the one of the same radios used by your phone, tablet and computer.

The more direct your connection to the Internet, the easier it is for hackers, bad actors, to break into your system and do bad things. Wi-Fi is a straight pipe to the Internet. One company that allows you to control their light bulbs using your smart phone over Wi-Fi was found to be doing careless things with your passwords, creating an opportunity for people to break into your computer.

Bluetooth

Bluetooth in its various forms is good for connecting your smart phone to things like ear phones, speakers and even toothbrushes. It supports stereo audio, two-way communications for headphones and is good for about 30 feet. It is also the radio used to connect your smart phone to your car.

X-10

X-10 is the original multi-vendor home automation standard. It is no longer used for new applications.

Z-Wave

Z-Wave has been around longer than Zigbee. (Zigbee is discussed later.) Typically, a Z-Wave system has a master controller that can provide connectivity to the Internet. This connection can be made using an Ethernet wired connection or Wi-Fi. The Z-Wave radios use mesh technology to go further. In a mesh system, multiple *hops* are used if a device cannot reach the master controller. Messages can be passed from the master to units in range of the radio and from them to more distant devices. A Z-Wave system can use multiple hops to reach the furthest sensors. Z-Wave devices must *pair* (register or sync) with the master controller, which in a system with multiple hops means the controller must be able to go *walk about* or operate on battery to move around the house for initial setup. The pairing operation cannot happen over multiple hops.

An informal survey of devices on the Lowes web site in 2015 showed about two out of three of the devices used in IoT applications were Z-Wave. Consumer comments and ratings were consistently more negative for Z-Wave than for Zigbee. The chief complaint was devices would suddenly become *un-paired* (loose their connection) with the controller. Since pairing is a single hop, manual process this becomes tedious and a real pain. This matches my own Z-Wave experience.

Security is also an issue with Z-Wave. It was an afterthought. As such, security adds a considerable delay in message transmission for multiple hops. And there is no mechanism for secure key exchange. This means anyone listening (snooping electronically) to Z-Wave pairing can now make themselves part of your network. This is a huge problem... trust me.

Zigbee

Zigbee is similar, but different, from Z-Wave. It is a different protocol than Z-Wave. It uses mesh networking like Z-Wave. It had security designed in from the start. It is much harder to crack. Turning security on does not slow communication over multiple hops. Devices must be paired with a central controller. That controller usually connects to the Internet using an Ethernet wired connection or Wi-Fi.

Zigbee offers significantly better security but it has a fatal flaw. There is no secure key exchange. There are excellent chips that are extremely low power. Messages can be secured. As good as it is, I will not use Zigbee for a door lock. A bad guy listening to any pairing operation (yes, they are infrequent) can break into my Zigbee network.

I do not see complaints about losing pairing like I see with Z-Wave. Of the two technologies, the newer Zigbee is much better.

Some power meters use a form of Zigbee. These add secure key exchange and remove my reservations about the protocol security.

Controllers

Before you begin to use IoT, you must have a solid internet connection. *[See the chapter on Internet.]* Without Internet, you really can't do much home automation. Many devices will also require Wi-Fi.

You can have an internet connected device such as a thermostat without much trouble. It will connect via Wi-Fi to the cloud and allow you to adjust it and control it from your smartphone or tablet. But to really get going with home automation, you need a *controller*. These are also called *gateways* or *bridges*. They connect the low power radios used in the fog to the Internet and to the cloud. There are different brands and the players change daily. Be prepared to do a lot of tinkering. This stuff is not ready for prime time. The do-it-yourself pieces are not turnkey.

You can get a turnkey system from most Service Providers such as AT&T, Comcast, Charter and Cox. Some home security companies are also getting into home automation.

If the words "Your router must be NAT compatible" and "Ports 443 and 80 must be open through the firewall", strikes fear and loathing in your heart, run away. Or be prepared to buy lots of beer for younger more tech savvy friends or neighbors. Or go turnkey with a Service Provider.

When you work with home automation, you first choose a radio. The choices were outlined previously. Most sensors need a low power radio like Z-Wave or Zigbee. I've used both and greatly prefer Zigbee. You will see things like HA (Home Automation) profile. This is a software set of handshakes. Devices that do the same handshake will work together better. Distance is a factor. Your controller must be connected via wired Ethernet to your router (internet connection). It may not be at the center of your house. To reach longer distances, the low power radios use *mesh*. The devices closer to the controller pass messages along to devices further away. This works well when you get it set up.

All this complexity and interoperation mean that you need to stick with a bigger player who offers most of what you need in a single package. Not the cheapest route perhaps, but the most trouble free.

Iris by Lowes

Lowes, the big home improvement chain, has an impressive amount of newer technology, from a wide selection of LED light bulbs to its own branded Home Automation kit, Iris. About two years ago, everything at Lowes revolved around Iris and being Iris compatible. Today they offer a bit broader selection, but Iris is still the primary IoT product family. You can get the controller as a standalone unit, or in a wide variety of kits to accomplish a set of tasks such as lighting control or home security.

Iris supports Zigbee HA (Home Automation), Z-Wave, Wi-Fi and Bluetooth radios. Iris requires a hardwired Ethernet connection to your router.

Customer reviews typically rate the Zigbee products as more trouble free than Z-Wave.

Amazon Echo

Another big name in the business is Amazon. At

this writing, they do not have a controller. But their Echo device can operate many controllers like Wink, Samsung, Insteon and Nest. It is an internet connected speaker that listens to your voice and will do things on request. As a speaker, it can play streaming music such as Amazon Prime Music, Spotify, Pandora and iHeart Radio. It can access web sites and read you information such as the weather or audio books. You can use voice control to drive other controllers to turn on lights, change the temperature, buy products on Amazon and even unlock doors.

Figure 36 Amazon Echo Wikipedia.org

Amazon is a big enough player with enough market power and cash to do many things. They are worth watching. Amazon Echo gets good reviews.

I said this is the wild west... be careful.

You get Amazon Echo's attention with a wake word, "Alexa". One young girl watched her parents ordering things from Amazon using the Echo and then said "Alexa, I want a doll house." The family was surprised when a couple of days later a large box arrived from Amazon containing a doll house. The little girl fessed up and the story made the local news.

Apparently when the story played on the local news channel, Amazon saw a huge spike in orders of doll houses from this one city. Listeners with Echo had the ordering process triggered by the sound from the TV News show.

The ordering feature is turned on by default, of course. But you can turn it off. And you can change the wake word "Alexa" to one of several others. I would recommend all the above.

There is another possible hack. If you can make any device in the home such as a TV Channel or audio enabled security camera say "Alexa, unlock the doors". Then it becomes easy for the bad guys to enter your house. Go slow with giving door access to IoT.

Google

Google bought Nest. Nest is best known for their thermostats but they also have smoke and CO_2 alarms. Google now has Google Home. It is like Echo but as a newer product has fewer connections to third parties. This will change with time.

Figure 37 Google Home Wikipedia.org

A CFO friend, Rick, asked, "Why did Google buy a hardware company like Nest?" My answer was "for the data." Nest knows when you are home, what your schedule is, when you go on vacation, what the weather is in your neighborhood. When you install Nest or other internet connected devices, your life becomes open to these large companies. Google is betting they can use your patterns to make money in advertising. Which is why Tom's wife would not let him install it. She was not willing to trade her privacy for energy savings.

Should I be concerned with Amazon Echo and Google Home recording me 24/7?

It depends... When you use web browsers like Internet Explorer, Safari, Chrome or others, it remembers your history and can remember passwords. It also stores cookies from the web sites you visit to remember things. This was all intended for the good, but at times it can be misused. Your voice and the other sounds in your home being recorded is new enough that it gives us pause. Is a stranger listening to my most intimate conversations?

Both the Echo and Google Home are smart, but not so smart that they understand all human speech and know the answer to most questions. This requires the big brain located in the cloud Amazon and Google both use wake words. The defaults are "Alexa" and "OK Google". They are designed to listen always but not transmit until they hear their wake word. Then what is picked up by the microphone is transmitted via the Internet to the smart computers in the cloud which figure out what you are talking about. These recordings are kept associated with your account and help train the system to respond to your voice and to help Amazon and Google to do a better job for all users.

You can use the respective app to listen to your interactions with your digital helpers and even delete some or all of them. But just like cleaning out your browser history and cookies, when you delete these utterances, you make the system dumber and less able to read your mind. You can also turn off the microphone.

Microsoft does not have a standalone voice enabled helper but Windows 10 on most computers does have Cortana which is similar. It is finer grained and works differently but it is closer in operation to Alexa and Google Home than Siri. While writing this book Microsoft has announced an intention to release a similar device. See www.simpleguide.tech for updates.

AT&T

AT&T is another big company. AT&T is losing POTS (plain old telephone service) customers at a rapid pace. They stopped the bleeding by buying a wireless (cellphone) company but to grow, they need new business. They are betting on Home Security and Home Automation. They also have added video with Direct TV.

Home Security today is a patchwork of small security companies. AT&T is betting that they can come in with their large size and dominate the market. Security Systems require a monitoring service. A monitoring service provides a human to apply judgment and pass the alarm over to the police in a timely fashion when they know it is not a false alarm. There is a monthly fee for this in any setting. The security portion of AT&T's system uses a highly secure proprietary radio. It does not have the weakness of Zigbee or Z-wave.

The AT&T controller has multiple paths to the Internet. The most common is wired such as DSL or optical fiber. It can also use the cell phone network, 4G to communicate.

In addition to the proprietary radio used for home security, AT&T can use either Z-Wave or Zigbee to perform other home automation functions. They tend to allow only certain devices and brands to connect to their controller.

AT&T is the Cadillac of the home security and home automation business. For that you get a reliable turnkey solution. And for that you pay Cadillac prices.

Cable Company

Cox, Comcast, Charter and other cable operators offer a similar home security, home automation product line. They use more off the shelf components and the offerings are constantly improving. Before you sign on with AT&T, you need to check them out.

Apple Home Kit

Apple at this writing does not have an Apple branded controller. They do have Siri. Using your Apple iPhone or iPad, you can ask Siri to do several things. To extend this to home automation Apple had to develop a common language. This is Apple Home Kit. If you are an Apple user, you want to make sure your chosen controller is Apple Home Kit compatible.

Caseta Wireless Smart Bridge

Caseta is made by Lutron. They are a lighting company. The Smart Bridge is a controller that, from the web site, seems pretty much limited to controlling Lutron lighting. It can be controlled by Amazon Echo or Apple Siri.

A limited package like this makes sense if you want to break lighting off into its own package, and use other controllers for other functions. I have no time with this product.

Wink

Wink is available from Home Depot and Amazon. They are independent. They have Wi-Fi, Bluetooth, Zigbee and Z-Wave radios. Their web site says they work with Google and Apple phones but not tablets. This will likely change in time. They do work with Nest and Amazon Echo. They do not currently support Apple Home kit.

I like Wink. They don't have a monthly charge like several other players. But you should wonder if they can get critical mass before being crushed by larger players. Today they are very credible if you have some tech savvy.

Samsung Smart Things Hub

Samsung is the big Korean appliance manufacturer. I am liking more and more stuff made by Samsung. Their entry into Home security and automation is very new. It was not on my radar 12 months ago. They are big and well-funded. They make great stuff. Basically, it has Wi-Fi, Zigbee and Z-Wave radios.

I think you need to keep an eye on this. It could be good stuff. If you are interested.

Logitech Harmony

Logitech makes great universal remotes which is a real challenging problem to solve. They are now trying to get into the Home automation market.

Home Metering

The folks that put power meters on the side of your home are tired of sending a person with a clipboard around to read your meter. This gets even more complicated when the Power Company wants to bill different rates at different times of the

Figure 38 Power Meter Wikipedia.org

day. Unless they can talk directly to that meter they cannot get the data. One technology for reading that meter is to use an internet connected Zigbee controller in your home that talks to the meter using a Zigbee radio. But this is your power bill. We are talking real money here so they must address that pesky *pairing* security hole. And they have. Power meters have added an extra layer of cryptographic communication to the already robust Zigbee standard. When pairing occurs, they use a different method to secure that handshake.

Applications

Home Security

When I built my home, my security contractor ran wires to every door and window and to many sensor locations. It was not cheap. If you have an existing home, it becomes even more expensive to run wire. In these applications, wireless radios can save the day. You need good security and long battery life. These remote sensors run on batteries. And because they may be distant from the main controller but close to each other, a mesh network is ideal. Zigbee fills the requirements nicely. If you can afford AT&T, they have a better proprietary radio. But you are more limited with sensors. The final security hole in Zigbee is slight. Your average home owner will be fine with Zigbee (Not the doors, please).

In addition to door and window sensors, you can add TV cameras, infrared sensors, garage door controllers, doorbell, 2-way video and other applications too numerous to mention here.

Heating and Air

If you look at the average household budget's monthly expenditures, a key opportunity for savings is in the Heating and Air Conditioning (HVAC) bill. Several companies offered *smart thermostats* before Nest came along. But truth be told, they were not easy to use and they were not all that smart. The Nest thermostat learns your patterns, knows the outside weather, learns the capacity of your HVAC unit and works to minimize what you pay to make your home comfortable. It connects to your HVAC unit and runs off the very low voltage on the control wires without a battery. Nest connects to the Internet using Wi-Fi. It is an admirable feat of design and engineering. Now that they have proven it can be done several mainline HVAC companies have controllers that do a pretty fair job on their own.

Still it is disconcerting to have a device learn my daily routine and report back to the mother ship. But I like the money it saves me.

Irrigation

Not everyone has a yard irrigation system. But if you drive down the street and see the sprinklers running on a rainy day you know it is an opportunity to do a better job. Optimal watering is an art. You water turf differently than the shrubs and flower beds. Watering

Figure 39 Lawn Sprinkler
Wikipedia.org

depends on the type of soil and the slope of the yard. The traditional units are dumb. And they are complicated. My friend Tom, built a new house and his landscaper installed the system and locked the door of the unit. Tom was told it was too complicated and that if he tried to change it, the unit would likely stop working. There is something wrong with this picture. Another friend, Al, tried to turn off one section and broke other things. It took Al hours of work to get the system back to some semblance of order.

I thought about building a better unit, but several companies were ahead of me in this endeavor. Yet a year later, nothing has taken off like the Nest. I don't know of a turnkey system I can recommend today. But there is opportunity here.

Fire and gas

Several companies including Nest now make
connected smoke, heat and CO2 detectors. The
Nest units use Wi-Fi. Other units use Zigbee and
Z-Wave.

Lighting

Most LED light bulbs are not internet connected.
The average LED bulb simply provides light.
There are a few internet connected bulbs that
allow you to dim the light or change the color of
the light using an app on your smart phone.

I would have thought that putting Wi-Fi in a light
bulb would not be cost effective. Perhaps it is
not. Putting in Wi-Fi or Zigbee into the light
socket would make more sense. It also takes
more work. Given that LED light bulbs last 20
years under normal use, maybe the cost is not so
bad. If you recall from the lighting chapter, LED
bulbs get white light from a combination of red,
green and blue LEDs. This allows them not only
to turn on and off, bright and dim, but also
change color by controlling each color of LED
separately. Given that the average socket is 120V
AC and was designed to house a very inefficient
incandescent bulb, power is no object. So, Wi-Fi
is the easy solution.

Lots of companies make these. Osram was one
of the first to offer control of color as well as
brightness. Lutron makes a wide variety of

products. Also look at who your controller partners with.

Osram light bulbs

A word of caution: Osram was just reported to have a major security flaw. Osram uses Wi-Fi rather than Zigbee or Z-Wave. Thus, their light bulbs are part of your home network. The same one used by your computer with your banking log-in, your 401k data and your family photos. Security researchers found that Osram was keeping your log-in credentials, the keys to your personal data kingdom, in an insecure manner on a database in the Cloud. There have been no reports of people losing their retirement savings by installing a light bulb yet. But you need to be aware of the risk you are taking putting IoT on your network. My advice, use Zigbee or Z-Wave to talk to IoT devices.

Phillips Hue

Canadian and Israeli researchers have found a security flaw in the Phillips Hue internet connected light bulbs that could theoretically be exploited in a hack. Theoretical exploits mean we don't know of an actual attack. But one might be possible

Locks

In case I wasn't clear earlier, I would not use a Z-Wave connected lock on a dog house. If I wanted badly enough to make my door locks part of my home network, I might use Zigbee. But I have not and have no plans to do this.

There are nice locks that use Bluetooth to let your phone open them. You can send a door code key to a friend's phone that would allow them to open the door.

I have locks that use key pads. I don't think traditional keys are the answer. Just be aware of the risks.

IoT Security

Security takes place on multiple levels. Is the exchange of information encrypted? Encryption requires that sender and receiver have keys. Can you exchange keys without bad guys listening in? We see that Zigbee is good, I give it a B. AT&T security sensors use a proprietary communication protocol with excellent encryption. Just what you expect from a security system from a top ranked player charging top dollar.

When you use your smart phone or tablet to control your thermostat, security system, water sensors, there is usually a web site in the cloud used to store information and communicate between your phone or tablet and the controller in the home. Occasionally, the phone can talk directly to the controller but there is almost always a web site in the cloud. If your data is not secured on that cloud link, you don't have real security. This is the problem faced by the IoT light bulb company. They stored passwords insecurely in the cloud.

Studies show that even large companies are likely to have their IoT networks broken into in the next few years. Until consumers are comfortable with the security of their systems, deployments are a bit of the wild west. The fastest gun wins. The potential for IoT is huge. Today we are in an early adopter phase.

Do you know how to spot the pioneers? They are the folks with arrows in their backs. Go slow, be skeptical. But do not ignore IoT. It will change all our lives... in time.

Chapter 8 Content on the Internet

I t is not called the World Wide Web for nothing. I can't begin to scratch the surface here. But it needs to be addressed. How do you find what you need on the Internet?

There once were companies called AOL and Yahoo. They were portals to content from other places and even had content on their own. If you wanted to find something you went to AOL and Yahoo and you likely had web based email provided by them. "You've got Mail". Fast forward a few years and more than a few twists and turns and now they are both owned by Verizon (a phone company) and are no longer the default location for finding information. They are but a shadow of their former selves.

Welcome to life on internet time.

Now when people want information they tend to "Google it." Google has become a verb. My friend Steve swears Microsoft Bing is better. Try them both and see what suits your fancy. Will Google and Bing survive? I don't know and don't care. If they die off it is because something better came along. That is why this book has the year so prominently in the title. Things change.

There are newer companies providing portals to content. I have not used them yet. Look at igHome, NetVibes, Start.me and YourPort. Give them a look.

Even with the decline of traditional portals, there are broad categories of content we can discuss.

What happened to World Book?

When I was coming up in the world, we used printed encyclopedias for information. The

standard family encyclopedia was World Book. If your family was middle class, you had a set for your kids. It was an investment

Figure 40 World Book Encyclopedia Wikipedia.org

in their intellectual development. My own family growing up, had Encyclopedia Britannica, much to my dismay until I was much older. It was the adult version. It was hard for a third grader to understand. All printed encyclopedias went obsolete very quickly. Every year you bought the annual appendix; all the new developments that were not in the original 26 volumes. Soon you had more almanacs than you had volumes A to Z. And the atlas of the world was sadly out of date. Still there was nothing better. My wife and I bought a set for our kids. Yes, it was a newer, current, Encyclopedia Britannica, but it came with a second set of books for younger children. No more World Book envy. But then computers happened.

My wife loves to tell the story about the time I spent my bonus on a computer rather than a new couch. She was opposed. "No one would have computers in their home." She did not believe that everyone would indeed have a computer one day soon. Microsoft, saw the flaw in the encyclopedia business. Without a yearly refresh, the information was soon out of date. And it was limited to printed words and photos. No audio files of Martin Luther King's "I have a dream" speech. No videos of the moon landings. Microsoft came up with a CD based encyclopedia, Encarta. You bought the annual subscription and your encyclopedia was always up to date. Encyclopedia Britannica and others tried to compete but it was too late. Before they got off the ground a new revolution in information was happening and all the CD based encyclopedias were extinct like the dinosaurs.

Fast forward to the real Internet. And the idea that lots of people that are experts in one or two areas would contribute knowledge for free for other people to read. All that was needed was a platform, a bit of software, and a home in the cloud. Wikipedia was born. The English version today has over 5,209,000 articles. Many are very technical and almost always up to date. It was a bit of a wild west for a while. People wrote untrue things, biased information. But other people disagreed and eventually rules and protocols were put in place to tame the worst excesses. Good articles have links to other articles and even to standards bodies. By and large, it is reliable. And it has killed the encyclopedia business as we once knew it. It even killed Encarta.

We eventually threw away the Encyclopedia book shelf. We kept it for longer than it deserved.

It is worth noting that I don't just happen to remember all the dates and bits of history I have included in this book. Much of the base material was taken from Wikipedia. And many of the images are also available in the commons and licensed for commercial re-use. There is a lot of good stuff there.

News

IMHO (In My Humble Opinion), current news deserves to be curated. I want someone to help me separate fact from fiction. And separate the trivial from the important. News is one thing I will pay for. As I re-edit this book again to get it ready for publication a new topic has popped up, Fake News. It turns out that you can make stories up and put them on a web site. If you can entice people to view your web site, advertisers will pay to place ads next to the phony articles. This goes a step further than choosing real news to get ratings but it is a slippery slope. Stick with the tried and true and keep a healthy skepticism about truth and bias.

Facebook

I have read that a significant number of people get their news from Facebook. This in my opinion is a sad commentary on society. *[I sound like an old geezer, don't I?]* FB is now the source of many links to Fake News. Just because an old high school friend reposts it, doesn't make it true.

New York Times

New York Times is available on the web. And they have a nice iPad app. My guess is you can get them on Google's Chrome as well. Pricy but well respected curator of the news. I can find all the same stuff for free. I just pay them to sort it for me. They do have a bit of bias but they try hard.

Wall Street Journal

Same thing. Well respected, Nice app. Good curators. More business and finance.

CNN

Free. Web based. Tends to more video. I like to read my news.

BBC News

I like different viewpoints. This one is UK based. You pick up different articles from around the world. BBC also has a nice podcast called News Hour. You can find the podcast by searching for

BBC News Hour podcast. It is updated twice a day.

USA Today

This is a web version of the free newspaper you get when you stay at a hotel anywhere in the country. It sometimes picks up breaking news faster than NYT or WSJ.

Reddit

Reddit is not news. I don't know what it is. It has a bunch of special interest groups. There is a group (fans) who talk about an old obsolete video game product my team made years ago. There is a good group on the Appalachian Trail, my son used frequently to plan his summer pilgrimage. Reddit has enough interesting groups with people that have specialized knowledge that you should check it out. But you may have to dig through the noise to find the nuggets.

Magazines

There are a bunch out there. Personally, I don't like looking at them on computer. I do find those that are optimized for my iPad to be good. This is how I consume my "Wired Magazine" subscription. They are hip and have lots of links to advertisers and expanded content. Harvard Business Review, Time and Sound and Vision are some of the names of magazines available on the web.

Talk to your local public library. Many libraries offer their members access to magazine subscriptions for free through the Zinio app. These subscriptions are expensive if you simply subscribe. It is a nice use of tax dollars to allow us to check them out on line. (Thanks for the PSA, John)

Look in your favorite magazines for an on-line option or go to your favorite search engine (such as Google or Bing) and type in "*topic* Magazine" where topic is the thing you like such as model railroads or cats. I'm sure you will find something.

As much as I may like reading a paper magazine, the cost of printing and distributing may force them all to go to the Internet.

IMDB (How to settle an argument)

You have friend over and you can't agree whether Mel Gibson ever worked with Cate Blanchett. Things may be heating up and you need to solve the problem once and for

Figure 41 IMDb Logo Wikipedia.org

all. IMDB or the Internet Movie Data Base is the golden reference for all things movie and TV. You're watching a movie with your spouse who says, "that actor is Bing Crosby." And you say,

"No it can't be." So, you search on the show and sure enough, that actor is Mickey Smith. Crisis averted.

By the way, you have likely heard of the six degrees of separation from Kevin Bacon. This was a major research paper done using the IMDB database. It turns out Kevin Bacon is widely popular and has worked on a lot of movies. He is also eclectic and works on many different types of movies. He works on movies with lots of different groups of people. Bottom line is that most people in the movie and TV business have only six or fewer degrees of separation from Kevin Bacon. Thank you IMDB for that tidbit.

Here is one that broke out recently in my family. What movie based on a John Grisham novel, did Danny Glover have a major but uncredited role?

YouTube

Figure 42 YouTube Logo
Wikipedia.org

YouTube is another Google company. It was known for cat videos when it got started. You can post almost any original content up there. There are entire Bollywood Movies on YouTube. But it can be a useful reference also. Our friend, Kathi, says anything I need to do, I can find how-to videos on YouTube.

When I wanted to scrape my popcorn ceiling. I looked up how to remove popcorn ceilings on YouTube. I found six or more videos. The first three told me how easy it was. So, I started the project. When it was not as easy as shown, I watched the fourth video. It told me the problem was that popcorn ceilings are hard to remove when they have been painted. (It advised selling the house) I've used YouTube to get information on topics as varied as using a new feature of Microsoft Excel spreadsheet or fixing a refrigerator.

You can find comedians there, singers and my son's sermons. In short, it is a vast cornucopia of video entertainment.

Traffic data taken off the internet backbone shows that Netflix takes about 60% of the bandwidth during prime time viewing. Followed closely by YouTube with 25%.

Chapter 9 Artificial Intelligence (A.I.)

I ntelligence: The ability to learn or understand things or to deal with new or difficult situations.

Artificial: not happening or existing naturally: created or caused by people

Artificial Intelligence: the power of a machine to copy human behavior, an area of computer science.

Merriam-Webster

> The key thing about A.I. is voice control for 2017. Alexia, OK, Google, Bixby, Siri, Cortana. And the nameless interface to your car.
>
> They are getting better. And they will be scary good soon.

Why have a chapter on AI?

Artificial Intelligence is a core technology that enables many of the products and technologies discussed in this book. A.I. is behind Google and Bing Searches. It's behind self-driving cars. It even drives what YouTube and Netflix offer up to you as suggestions for what to watch. And it will get better very fast. It is behind the move to voice-enabled interfaces. It will affect every part of modern life, dizzying as that may seem.

What is it?

If you want to start an argument among computer scientists, ask the question "What is artificial intelligence?" You recall the old story of the four blind men and the elephant? The one blind man grabs the elephant's trunk and declares that the beast is like a snake. Another grabs the leg and says it is like a tree. Another touches the side of the elephant and says it is like a wall. The fourth blind man grabs the tail and says, "You are all wrong. It is like a rope.". Artificial Intelligence is many things, all correct, depending on which part of the beast you touch.

Rather than define it, let's talk about how it is sneaking into our daily lives. And remember, Artificial Intelligence is no match for human stupidity. (Thanks Ivan)

Thermostats

The first thermostats were *bang-bang* or on/off. You set a temperature and if the temperature falls below the setting, the thermostat turns the heat on. When the temperature reaches that setting, the thermostat turns the heat off. This results in very frequent on/off cycles and is hard on the equipment.

This behavior was modified later to make it turn on at one temperature and off at a different temperature.

Figure 43 Nest Thermostat
Wikipedia.org

The heat did not turn on until the temperature dropped X degrees below the target temperature. And it turned off when it reached the target temperature.

Today, the best thermostats learn the patterns of the people inhabiting the space. When are they home? When do they adjust the temperature? How does the equipment behave? How long does it take to heat or cool the space? What is the temperature outside? What is the cost of electricity at this time of day? Do you prefer to save money or be comfortable? This ability to learn the behaviors of people and the machinery it controls puts the new thermostats like Google's Nest firmly into the A.I. camp. Others will find the use of big data, such as the weather forecast, to be evidence of A.I. Either way, they have gotten pretty smart and I know that the algorithms (rules) that control my Nest thermostat have been updated more than once while I owned it.

Sometimes our devices get too smart and do strange things. I find my current Nest thermostat is adding temperature changes during the day or week that I did not set up. I must pull up the controls on my tablet device and delete the extra temperature changes every few months. I hate it when a device thinks they are smarter than I am. Especially when they are correct.

Voice recognition

Our 2005 Acura cars had voice recognition. My wife never used it. I had minor success with it but it was limited. My 2015 Ford truck has much

For those of you that experienced the early days of computers trying to understand our language, you may recall a game called 'Adventure'. It did not listen to your talking but it did try to understand sentences that you typed in. We quickly adapted to its need for a verb and noun.

The game would describe a scene and you would respond with an action. "You are standing in a cave. There is a troll in the cave. There is an axe on the floor." You replied, "Kill the troll". The computer would say "You have no weapon". You could say "Pick up the ax" and the computer might say "You cannot carry any more". To be successful you would drop the gold, pick up the ax and kill the troll. Then you dropped the ax and picked up the gold before the pirate appeared to steal your valuables.

It was a slower pace than today's computer games. My son would hate it.

better voice recognition and after Ford switched to a faster computer with more memory in 2016 (Sync 2) it has gotten better and better. Expect rapid advances in being able to talk to your stuff.

For reference, the early speech recognition had to be trained to a specific speaker. Today, the goal is to understand anyone with a wide variety of accents and no training. The more limited the possibilities, the better the system performs.

General Purpose: Apple Siri, Amazon Alexa, Google Home

Siri came along on the iPhone and immediately became the inspiration for late night comics and screen writers everywhere. Unlike your car, Siri uses computer power in the cloud to figure out what you are saying and find the answer you seek. Press the home key (the dimple at the bottom of the screen) and ask a question like "What is the weather?" or "Find a coffee shop." You can also ask her if you are handsome but be prepared for some witty remark in return.

You can now have Siri on your iPhone, iPad, Apple computer with Mac-OS Sierra. Cars with iPhone integration also use Siri. Siri now has Apple's Home kit to connect to devices in the home for Internet of Things (IoT).

Amazon introduced the Echo as a Wi-Fi connected speaker with Alexa as their voice recognition device since Amazon does not have a phone. It allows you to stream music from the Internet using music services like Amazon Prime. It also connects with a wide variety of IoT devices. If you don't want to listen to music, there is the less expensive Echo Dot that does not have audio quality speakers.

These services assume a connection to the cloud.

Figure 44 Amazon Dot Fliker.com

If you have no internet connection they do not work. As we said earlier, they are not supposed to transmit sounds picked up by their microphones until they hear the wake word. But they are connected to the Internet. And all the above can be hacked.

Just don't say I did not warn you. Tin hats anyone?

Cars

Cars do not assume an internet connection. Their conversations are more constrained. If I ask Siri "Why is there air?" she/it pulls up a Wikipedia reference to the Bill Crosby 1965 comedy routine. (The answer is to fill volleyballs) If I ask Ford Sync the same question it says "I don't understand. Say the name of a device like Phone or Guidance or Radio". Without Internet, voice recognition is limited to well-known vocabulary. Current systems work without training, unlike my older cars.

To make cars and other *unconnected* devices work, the computer expects a limited vocabulary with no unexpected commentary. You are well off to learn the words it knows and stick to them. Who is smarter now? The computer is training us rather than we are training it.

Video

Apple TV Gen 3, Roku, Comcast Xfinity, Cox Contour 2 and assorted TVs will now listen to you. For the TV, this is a real trick because of the distance from the screen and the fact that TV shows come with audio. The Apple and Roku devices put the microphone in the remote. The Comcast and Cox apps are cool, you can say a famous line from a movie and if it is available for download the app pulls it up. Try "I'll be back" and you get Terminator. Or "Do I feel lucky?" Well, do ya, punk?" and you get Dirty Harry. It's fun. Amaze your friends.

I haven't played with my Apple TV or Roku but I know the TVs have a setting deep in the menu that allow you to turn off the always listening feature. Did I mention tin hats?

Facial Recognition Software

One of the neat tricks that Apple Photo, Facebook and others have mastered is facial recognition. I think Facebook does it to pull

The following tidbit is geeky but fun. The standard test for goodness in facial recognition is called "Labeled Faces in the Wild" or LFW. It has 13,000 photos of 5,000 people. And most companies score well above 95%. Google is almost perfect. But that is not the world we live in. It may be good at sorting out the limited faces in your photo album but not good for catching the criminal mastermind in video surveillance. A new test has been created called the Mega Face Challenge. It has 1,000,000 photos tagged in Flicker with a creative commons license. (The same system I used for illustrations in this book) of 690,000 people. In this test, Google scored only 75% and they won the bake off. There is still room for improvement.

While it works OK in Apple Photo, the biggest challenge for most of these systems is to match faces from different ages.

people in and increase usage. Photo does it to help you categorize the zillions of photos you snap with your phone. You can now sort by location based on the phone's location information being tagged to a photo, time and date, who is in the picture based on facial recognition, and any other collection or tag the user takes the time to apply.

When you take a digital picture, nearly all devices will add meta-data. Meta-data is data about the data, i.e. photograph. It can include time, date and settings used by the camera. If your camera has a GPS receiver, it can store location information. This would allow someone with access to the picture to know more than perhaps you intended. I would recommend stripping out meta-data before sharing a photo on line.

In Apple Photo, when you export a photo, uncheck both options in the info section. In Windows 10, open the folder containing your photos, select the photos you want to share. Right click and choose properties. Click details and select Remove properties and personal information.

Self-Driving Cars

We remember the cartoon The Jetsons, and we all thought by the time we were grown that cars

Back to the future. AT&T Picture Phones arrived 50 years after they were presented. Self-driving cars were announced in 1960 by Radio Corporation of America (RCA) and General Motors (GM). These cars followed a wire embedded in the road and were predicted to be commonplace by 1975. It's hard to predict the future.

In 2013, South Korea was trying to adopt the follow-the-wire technology to buses. Today's self-driving cars use LIDAR (a radar like system based on lasers) computer vision and GPS with lots of computer power.

would drive themselves. We are getting there finally if you ignore the flying part. Maybe? Soon? Of course, by the time this is published it will be out of date. Check the web site for the updates.

A self-driving car is made possible by lots of different technology coming together. You have GPS to know where you are within a few feet. You have sensors that warn you about items nearby. Cameras see the road ahead and behind. Computers that process this information and act or inform the person behind the wheel. Is it ready for prime time? Not for me.

Uber is testing in Pittsburgh. They have a driver behind the wheel. Google has cars roaming California and other states where the laws allow. They have operators to take control if the car does not behave. Usually the problem is that the car won't go because it cannot guarantee safety. The problem is the chaotic environment. The same thing that gives human drivers fits today. Kids running out between cars to chase a ball. Drivers that don't yield at four-way stops. Lack of communication between cars. Poor signage and poor maps in the boondocks. There is a proposal to make semi-trucks autonomous. Less chaos on the interstate than in a city. But given the size, speed and mass of these vehicles, it scares the living fire out of me.

Several teams of self-driving car developers were asked when the technology would be available in India. The answer was "Not for a very long time". Having worked in India, I can attest to the chaotic nature of the traffic there.

Regardless of success on self-driving cars, all this technology has one major payoff. It makes you safer. Ford has talked about their go-slow approach to the topic. Their focus is less on taking the driver out of the loop and more on helping the driver avoid serious mistakes. This is true today, but in the months, it has taken to produce this book, Ford has bought a company working on self-driving cars and is projecting releasing a real model in the early 2020s.

Directions

Mapping was an early beneficiary of big data and fast computers all residing in the cloud. You type in an address and a map pops up showing the address. Click on directions and it asks for a second address. Put in the second address and it gives you a list of directions to get there. Often it will give you multiple routes. Click on a button labeled, nearby, and it asks if you want hotels, restaurants or parks.

Appliances

One of the subtlest beneficiaries of A.I. is household appliances.

Vacuum cleaners

Roomba and others can do a fair job of picking up the day to day light dust and dirt. It is mobile and thus benefits from battery power. Because of the limits of battery power, the amount of sheer horsepower to suck up dirt is limited. What is

Figure 45 Roomba vacuum cleaner Wikipedia.org

interesting is watching the circular disk run about the floor in patterns designed to avoid obstacles and still clean the whole floor. When the battery runs low, the Roomba remembers where the charging station is and will go recharge its battery. Roomba makes versions that mop floors and even a gutter cleaner. Other companies now compete in a market that Roomba dominated for a time.

Ovens

Ovens and microwaves are starting to ship with sensors that can tell how warm a dish is and adjust the cooking. I haven't bought an oven in many years so I have no experience in this category.

Clothes washers

Clothes washers are getting smarter. They sense the weight of clothes and how dirty the water is and will adjust their washing. If you told your smart washer to do a normal load for 40 minutes and came back 20 minutes later to find 30 minutes on the timer, the washing machine will have adjusted its cycle based on what it senses in the load.

Robots

Big industrial robots used to be dumb machines. They only did what you told them to do. The programmer had to write the detailed instructions and put a fence around the device to keep it from whacking people that got in its way. This has all changed. Robots today have computer vision, sensors that can tell how hard the hand is gripping and can sense when people get in the way.

> One robot in a university lab was programmed to learn by watching people perform a task. The researchers were surprised when the robot was placing a nut on a screw and started by turning it in reverse before screwing it on.
>
> The robot had observed the humans do this and incorporated it into its learning. It helps to properly seat the nut on the screw and avoids cross threading and damaging the screw threads.

In Japan, Sony made robotic dogs that were kept as pets. They are now obsolete but continue to be very popular. There is a great secondary market in repairing and keeping these robotic pets running.

Also in Japan, they have robots that look like people to answer question at airports and check people in to hotels. You usually have a choice to check in with a robot or a human.

Some companies use robots to deliver mail. Hotels are experimenting with room service and companies like Amazon are using more and more robots in the warehouse. Whatever the application, expect to see more and more robotic helpers in your life time. But don't expect them to look like Rosie from the Jetsons.

Chapter 10 The Junk Drawer

We all have one. That drawer that we throw things into when we don't have a place for it. So why not this book? These are short miscellaneous topics that you might have questions about. And in the spirit of being the engineer next door you can ask questions of, I'll try to anticipate your

Figure 46 Junk Drawer Flicker.com

questions and comment. There is an email Questions@suwaneepress.com if you have more questions you want answered in the updated volume. Who knows, I might even answer you directly if I can.

There is also a blog on related topics at simpleguide.tech.

Bitcoin

What is Bitcoin? It is a new currency. My dear mother-in-law says that "money is just a medium of exchange." Usually she says this when she has spent a bundle on a vacation for her large family or big presents for the grandkids. Bitcoin is a medium of exchange. You can, in certain places, buy things with it. It is not backed by a government. It fluctuates wildly. As I write, it is worth about $1,076. This is close to a record high. When the book started, it was down around $500. It started close to zero in 2012. Volatility is seven times the price of gold, eight times the S&P500 and eighteen times the US Dollar according to Mark Williams.

Bitcoins are based on encryption and public ledgers called block-chains. The encryption involves public and private Keys. This is the same technology Pretty Good Privacy uses for email. Yes, dear reader we are deep into geek stuff here. If you lose your private key, you lose your bit coins. If someone steals your private key, they steal your Bitcoins. Remember they are almost $1,000 each. Thefts and hacks are common. A recent theft involved over $60M dollars.

Why am I talking about Bitcoins? Well, because your inquiring minds want to know. Bitcoins are interesting. Banks are trying to figure out how to use similar technology in future bank transactions.

If words like public/private key encryption and block-chain make your eyes roll back in your head, bitcoin is not for you. There is great volatility. There is significant risk. Hang on to your Social Security check and stick with the S&P 500 index funds. Or gold, you could be better off with gold. Ignore those articles about the guy who bought bitcoins in 2013 and now is a millionaire. It is not 2013 anymore.

Z-Cash

Bitcoins can be traced. There is a degree of anonymity, they are accepted on the *dark web* to buy illegal items like drugs and your rogue nuclear weapon. But they are not anonymous. They can be traced with some work. There is a new development called Z-Cash. Z-Cash uses the same block-chain technology, but the coins do not retain a history of trading. This imitates real cash. Z-Cash is not yet accepted at your neighborhood 7-11 but keep an eye out.

Block-Chain

You are going to hear this phrase in the future. It can be used for more than just money. It is based on cryptography. Fundamentally it can be used to verify and trace transactions, any kind of transactions. For example: Where do the ingredients of my salad come from? This can be important if you find that a lettuce grown in a certain field has salmonella. You then know which packages to recall. You can also trace things like electrical components to help find which products were built with a faulty batch.

Block-chains are very new. IBM has invested heavily in applying them to things outside virtual currency. If you are a CEO or in charge of quality control for a company you should consider this.

Texting? Email? Lives forever.

OK, I think it was Ann Landers or the modern-

day equivalent who said, "If you don't want to see it on the cover of the New York Times, don't put it in an email." This was certainly true in the 2016 elections. Our electronic servants are great at trying to remember stuff to *help*

Figure 47 Email Icon Wikipedia.org

us. Copies of photos, emails and documents get stashed in the Cloud, on backup drives and in various places on some not so volatile computer memory. Look at a gizmo on Amazon and ads for it show up on Facebook. Get a Gmail account and use Google Maps for directions and suddenly Google starts sending you notices of traffic delays along the way. This is useful, but unnerving at times. A quick photo with your phone suddenly shows up on your computer, your tablet and a copy in the cloud. If that is what you want, then all is good. Just be aware that with all this syncing going on that our photos and thoughts get lots of copies and are hard to purge if desired.

If you are using email at work, your employer has the right to read it. I was discussing a legal case with our company lawyers one time and I asked if they needed me to send them copies of files on my computer. Their answer was "No, we already took them from your backups." I was OK with this but shocked that everything on my hard drive was visible to the legal department and others.

I got the bright idea of buying a very large hard drive with *cloud syncing* for my wife and my self. Cloud syncing was a misnomer. It did not use the cloud as we described earlier. It did allow us to back up over our home's Wi-Fi. Even though I set up two different spaces for two different computers within a week the drive had mingled her files with my files. Her computer did not have a large hard drive to begin with and adding my files to it brought it to its knees. She did not need my unfinished computer programs. I did not need her recipes. I finally got the whole thing disconnected and started to untangle the files. It was very costly in time.

For file backups, I've had good luck with Apple's Time Machine. I've tried to copy key files as a backup to a backup. This is easy for spreadsheets, documents, photos and memory. But other things like email and checkbook ledgers are stashed in some unexpected locations that can change from version to version of your computer's software. Sometimes a backup doesn't pick up the nuances of licensed software. Or the latest device drivers.

One method that avoids all this is the use of clone software. When you clone a hard drive, you make a bit for bit mirror image. You can make a clone of your machine to an external drive and later if your hard drive crashes, you can simply insert the new drive and keep on working. But you lose all the work between when you made the copy and when you want to use the drive. Search for your operating system (Windows or OSX) and 'hard drive cloning' for free or commercial software. I've found the free stuff works fine for simple tasks. The commercial stuff is usually for big company IT pros.

Car hack. Key fobs, VW. Others

Just when cars are harder to break into, we get enamored with new technology that makes it easy to unlock and start cars, the key fob. On newer cars having the key fob in your pants pocket or purse allows you to walk up to the card and by touching the door handle, unlock the door. Pressing the

Figure 48 Car Key Fob

'Start' button starts the car and allows you to drive away. But this ease of use has made it easier for the bad guys to steal from you. These hacks, computer based attacks, are based on three methods that I am aware of.

The chips in these key fobs come from a few manufacturers and so a weakness in one may be present across brands. I had an arm load of packages and pressed the 'open tail gate' button on my Ford truck. The trunk of the Japanese car next to mine opened along with the tailgate to my truck. I quickly closed the trunk of the other car and placed my packages in my truck bed and left. But it surprised me just a bit.

On some brands of Key Fobs (VW is mentioned), researchers listened to less than a dozen unlock commands from a key fob to a car and could break the code and unlock the car with a device costing less than $40. This was a computer-based mathematical attack. (cryptographic) A better encryption scheme from the manufacturer is the only fix.

The second hack involves extending the range of the radio signals with a specialized piece of hardware. Police and researchers in Europe have seen examples of RF (radio frequency) extenders used to break into or steal some very expensive cars. In this case, they have a dual radio system than can pick up the signal from a distant key (on the dresser beside your bed) and amplify it making it seem like it is next to the car. You can beat this attack by keeping your key in a metal box (cough drops or Altoids) or a specially made bag that shields RF devices such as "Fob Guard" or "Silent Pocket". Some people put them in the refrigerator but cold is bad for batteries. I don't recommend the refrigerator for the sake of the batteries.

The third hack also uses specialized hardware. Your garage door opener and your car key fob use a technique for generating commands called *rolling keys.* Each time a signal is sent it is different. Once a code is heard by the receiver, it cannot be used again for a long time. If you can use math to break the secret sequence, you can generate a new code. Or you can do something easier. You can hide a device under the car that receives the code and blocks the door or gate from hearing it. When the door does not open, the owner pushes the button again generating a new code. The pirate device then sends the first code, unlocking the door and saving the new code for later breaking and entering. A good rolling code system has a time out mechanism that can defeat this. But not all cars or garage doors have implemented it.

What have we learned from all this? Certainly, new technology implemented in a hurry can be worse than what it replaces.

Internet Connected Children's Toys

I feel sad that there is even a reason to write this paragraph. It could be considered magical for a child to carry on an actual conversation with an inanimate object like Barbie or a cute stuffed Teddy Bear. But there are two threats posed by these types of toys. The first is hacking. This is where bad people break into and misuse internet connected devices. If a toy has a microphone or a camera, then bad actors can break into and spy on your child. This frankly is very creepy.

Perhaps the bigger threat is the misuse of the data by the company that made the toy. I'm quite sure that Mattel has too much to lose to blatantly disregard children's privacy, but "Hello Barbie" or its cloud database could be subject to hacking. It is worth noting that Mattel does not attempt to sell "Hello Barbie" in Germany where privacy laws are much stricter.

Germany has banned the doll Cayla. Cayla and i-Que are made by Genesis Toys. Cayla records the child's voice to be interpreted in the cloud. This recording is used to respond to the child. However, this data is passed on to Nuance Communications. Norway cites concerns because Genesis Toys reserves the right to use the information to market to the child. This violates numerous EU privacy protections. These products are also under investigation in the US.

Security researchers have also found that Cloud Pets made by Spiral Toys has an unsecured database of messages affecting over 800,000 toys. Many accounts had no password or an easily cracked password leaving these voice messages open for abuse. Weak and missing passwords can be blamed on the user. But given the market, grandparents wanting to leave messages for the child, things can be improved to make it easier to implement good password policy.

I'm going to sound like a luddite here when I say, "Maybe all this technology is not something you want for your kids." I think there is merit to giving the kid a dumb stuffed Teddy Bear and allowing the child to use his imagination.

Let the buyer beware. 'Nuff said.

The Picture Phone

At the 1964 World's Fair in New York, AT&T

 introduced the Video phone. You could stand in line and have a 10-minute conversation with a complete stranger at Disneyland in California. The product vision was great even if the picture quality wasn't. And then, nothing happened. The problem

Figure 49 AT&T Picture Phone 1964 Wikipedia.org

was bandwidth. Video requires lots of bandwidth.

Twisted pair networks designed for voice just don't have the bandwidth.

Fast forward to the Internet and now Skype, Google+ and FaceTime offer free video two-way communications using your PC, tablet or smart phone if you have a broadband high speed internet connection.

It's a great way to connect to family members and even to read stories to the grandkids. This one is worth figuring out. We love interacting with our grandson, Josiah, on a weekly basis.

GPS

Global Positioning Service (GPS) or what happened to paper maps? Believe it or not, many

 good things come from the government and the military. Teflon came from the space program. The Internet was a DARPA (Defense Advance Research

Figure 50 Road Map
Wikipedia.org

Project) and GPS came from a desire to put a bomb on target. Previously, if you wanted to take out a bridge you flew over and dropped a bunch of bombs hoping one would hit it. The old shotgun approach. Between the plane and the bridge there was wind and other things that could make it easy to miss. You could of course increase your odds by flying lower. Closer to the guys shooting back at you. You improved their odds also. Or use more planes, more targets for the guys on the ground. But what if the bomb knew where it was supposed to go and simply went there on its own. What if it was smart enough? What the bomb needed was a sense of where it was.

So those crazy scientists invented a bunch of satellites. They circle the globe in a precise pattern and send out timing signals. And if a radio receiver compares the delay in those signals and does the math, it knows where on earth it is very precisely. Now the bomb can compare where it is to where it wants to be and steer itself to the target. This helps many systems such as cruise missiles, and drones. Stick it in a jeep or tank and your troops don't wander around lost. Hey, you could put this in a car.

And thus, the GPS industry was born. Everyone wants to get into the game, The Russians have one and the Chinese want their own. But if you know the math, you can listen to everyone and just like drawing a line, the more data points the more accurate you become.

You still need to solve the map problem. You need good maps. In South Korea, these are state secrets so GPS has little value in South Korea. But in the US, maps are good and getting better. There is also the meta-data issue. Meta data is data about data. For example; Where is the nearest gas station or Mexican restaurant? How do I get from here to the nearest Starbucks? By the way, avoid freeways and toll roads. Suddenly its complicated. The main players were Garmin and TomTom. Now Google and Apple have taken the lead.

I prefer my GPS to be part of the car. But if you are in an older car or rental car, your smart phone is the way to go. Driving with a smart phone in your hand is dangerous. Get a holder that sticks to the windshield or AC vents and program it while the car is in park. And have the voice to tell you when to turn.

Solar Panels / power

I think I have said before, "How do you know who the pioneers are? They are the ones with arrows in their backs!" Solar Power is a bit of the Wild West. It holds great promise and some immediate benefits but things are still evolving. Let's talk about some of the issues.

Figure 51 Solar Panels Wikipedia.org

Semiconductors are amazing things. Take an LED, you put a voltage across it and it produces light. For a solar cell, you hit it with light, and you get a voltage. Mass production has helped make solar power affordable. But, you need electric lights at night when there is no sun. And there are cloudy days. This means you need some way of storing the power you make when the sun shines. Many systems use batteries. Tesla, the electric car company, which also needs lots of big batteries, is building a large battery factory. These batteries will be used not only in cars but they are hoping to sell them to home owners with solar panels. Anything Tesla can do to increase the number of batteries sold will help them drive the cost of the battery lower.

Another strategy is to sell excess electricity to the power company and buy power from them when your solar panels are not enough. Power companies generally don't like this idea for a variety of reasons, but regulators like the idea of being *green* and avoiding global warming. The regulators have said to the power companies, "Suck it up, and make it happen." This is fine when there are only a few people. When the number of people pushing excess electricity on to the power grid gets large, so do all the problems. The power companies push back with compelling data. And the regulators decrease what the power company must pay the homeowner for his solar power. What was a good deal for the people with solar panels on their roof, suddenly becomes less financially attractive. And this is what is happening today. Expect more of it for the reasons I will discuss below.

Power companies in general built networks designed to take power from a big power plant and deliver it over a wide area. This is a great simplification because there is more than one power plant and networks connect to each other. Ignore the complexity for now. The power plant is the base of the tree feeding all the branches. If you add solar panels out on all the branches, suddenly you have power flowing this way and that. Load balancing (matching power creation to power consumption) starts to get tricky. The capacity of the network needs to change. This requires engineering, construction and capital expense. And the regulators are forcing the power company to buy up the extra power. The power company becomes the battery. If the power company can buy it cheap enough, it works. If the cost and added expense don't equal their other power sources like nuclear, coal and gas then the power company starts losing money. Oh wait, power companies are regulated utilities, they are guaranteed to make money. What happens is everybody's power bill goes up. And the regulators change the amount the power company pays for solar and what was once a reasonable investment becomes less so to the home owners that put in solar panels.

The other problem with solar is time of day. Consumers don't use power at the same rate all day long. They use very little from midnight to 6 AM. They use a lot on hot summer afternoons. They use a bunch in the early evening for washing and drying clothes, cooking dinner, washing the dishes, bathing the kids. Power companies don't have big batteries, at least the way you think about them. And if consumers want power, they are required to generate it. On demand. That is the rule they live by. (I know of one plant in South Carolina that pumps water from a river up into a lake at night using extra power. And during high demand, they let the water flow back into the river through turbine generators to make extra power.) The cheapest power is from large plants like a big nuclear plant. They produce power at a constant rate 24x7. It is hard to turn these plants on and off. To get extra power, oil and gas generators can be more easily turned on and off, up and down. But the fuel cost is higher. Add to this solar with time of day and weather issues and you start to see the headache from the power companies point of view.

Wind energy has some of the same issues. You get power only when the wind blows. But it is not tied to the day time hours. Fun Fact: Texas, which is thought of as an oil producer is one of the leading producers of wind power.

So what? If you live off the grid, get yourself some batteries and buy solar. But, if you are planning to use the local power company as your *battery*, be aware that the price the power company pays you today for your excess power, will likely be lower tomorrow. If it is an excellent deal today, then all your neighbors will pile on. This adds cost to the power company and it will succeed in negotiating a lower payback tomorrow. If you don't care about the cost, and the most important thing to you is global warming and being green then have a ball, buy solar. If you are counting the nickels in your monthly budget, look at trends. A good deal today, maybe less so tomorrow.

Small Solar Systems

I helped a neighbor, Johnathan, size a solar/battery unit for a small device that he wanted to install in a field. In doing so, I learned alot. The *10W system* advertised on Amazon had a 1W solar panel and he was not happy that it was not working. We ended up using a 10W solar panel set up at 60 degrees and pointed south. In the southeastern US, this will generate 8 W for about 5 hours a day. (8W x 5Hrs = 40Whrs) We fed the power through a controller to charge a battery and powered his device from the battery.

Lesson #1: Without the controller, the solar panel can consume power from the battery at night. This reduces the effectiveness of the system.

Lesson #2: A 10W panel on average produces 8W because as the sun moves, the panel is not pointed directly at the sun for most of the day.

Lesson #3: You are only producing power for about 4.5 to 5 hours per day.

Virtual Reality

Dictionary.com defines Virtual as "temporarily simulated or extended by computer software."

Figure 52 VR glasses
Pixabay.com

Virtual reality is a reality that is extended by computer software. Not helpful? Remember stereo photographs? You look through a device and the left eye saw one picture and the right eye saw a slightly different picture and your brain saw a three-dimensional picture? What if the images were not photographs but were very realistic computer generated pictures?

Imagine if the glasses you are wearing can sense the movement of your head, left/right, up/down. The image changes like you were looking at real life. Move forward and you move into the picture. Add 5.1 sound and you are in the middle of a computer game or a CT scan.

As of the time I'm writing this, the challenge is getting the headsets that can present two high resolution images with sensors that detect eye movement and head movement cheap enough and light enough. Advances in computer technology make this inevitable. When they start selling millions of them the price drops. And the number of applications for them explode. It becomes a cycle that feeds on itself. I don't know what the killer app will be, but there will be at least one.

PDAs

I did not intend to include
Personal Digital Assistants
(PDA) in my book. But I was
reading a proposed
government standard on
Emergency Alert Systems and
they defined PDA. Really?
That is so... '80s. I think PDAs
went the way of Donna
Summers and disco music.

*Figure 53 Palm Pilot
Wikipedia.org*

To be fair, the work of a PDA , storing calendars,
notes, address book and other information, now
resides in a Smart Phone. In fact, the Smart
Phone has consumed the camera, phone, PDA,
and music player all in one nifty device. When
you see "PDA" think Smart Phone.

Only the U.S. government is using the term PDA,
a decade after Apple introduced the iPhone.

Batteries

Batteries are in the news lately with Samsung's
Galaxy Note 7 cordless phone smoking and
catching fire prompting two major recalls. There
are several types with different tradeoffs.
Typically, you need to use the same type that
came with the device. Devices that re-charge
their batteries are usually designed to work with
a specific battery type. Increasingly, devices do
not have replaceable batteries because some

cheap third party batteries were not well made and can damage the device or catch fire. If your laptop catches fire on an airplane then the news article reads "Brand X Laptop catches fire". It does not read, "Mr. Wang's discount battery in a brand X laptop catches fire".

Zinc-Carbon

A Zinc-Carbon battery is a *primary* battery. Meaning it cannot be recharged. You use it and throw it away. It does not hold as much power as an alkaline battery but it is cheap. It comes in various sizes like AAA, AA, C, D.

Alkaline

Figure 54 Batteries
Wikipedia.org

Alkaline batteries are primary batteries. They are meant to be used and thrown away. They hold more power than Zinc-Carbon and have a better shelf life. They come in a wide variety of sizes.

Lead-acid

Lead-acid batteries are *secondary batteries*. This means they can be re-charged. They are more expensive to buy but because of re-use, they are cheaper over the life time of the battery. There are different types of lead-acid batteries. The liquid cell was the common car battery of my youth. You had to monitor the water in the battery and refill it with distilled water if it ran low. They had to be stored upright to avoid spilling the water and battery acid inside.

Newer Lead-Acid batteries are sealed. These are sometimes called maintenance free because you don't typically need to add water. And in this category, there are the Gel and AGM (Absorption Glass Matt). In the Gel battery, the liquid is in gel form and less likely to leak. The AGM battery contains the liquid in a thin material between the battery plates.

All lead-acid batteries should be stored fully charged. Storage in the discharged state will damage the battery. Your typical car battery is used to start the car and to maintain power to devices while the car is off. If you have a car or camper that is not used often, you should invest in a *trickle charger* to keep the battery charged. These can be plugged into a household power outlet or you can use a solar cell.

Nickel-cadmium

Nickel-cadmium or Ni-Cad batteries were more common a decade ago. They are secondary, meaning re-chargeable. They have some severe limitations. These batteries have *memory*. If you only discharge them by 10% repeatedly, they forget that they can store more power and will only deliver the 10%. To get the best life from these batteries you should always use them until they are drained and then recharge.

Nickle metal hydride

Nickle metal hydride (Ni-MH) are a secondary battery or rechargeable battery. They do not have the memory effect of the Ni-Cad batteries. They are replacing Ni-Cad in most applications but the chargers are slightly different. Check the documents to make sure that either battery is supported before swapping them out.

Lithium-ion

This is the new darling of the rechargeable market. Lithium-ion is a secondary battery. It is light and holds much more power than other rechargeable batteries. This is its strength and its weakness. If a Li-ion battery is poorly made or damaged, the energy contained in it can make it explode and catch fire. Airlines are not accepting shipments with larger quantities of Li-ion batteries. And this is battery type responsible for the Samsung Galaxy Note 7 problems.

My best advice for avoiding problems is stick with big name brands. They have incentive to avoid costly legal suites if the batteries catch on fire.

Chapter 11 Crystal Ball

This is the part where I give you my guess as to what is coming soon. And I do mean guess. Let me give you some horrid guesses.

Fig
Wikipedia.org

IBM's President, Thomas J. Watson, once said in 1943, **"I think there is a world market for five computers"**

This quote is widely repeated but not documented very well.

But it sounds good.

> *In 1980, IBM was building the first IBM PC in Boca Raton, FL. I was working as a field engineer for Intel and working to get our computer chip chosen as the brain of the PC. The higher ups at IBM thought of this new thing as a toy with a limited market. Rather than build it in an IBM factory they subcontracted to a firm in Huntsville, Alabama. Projections for the first year were 250,000 machines. IBM was surprised to sell that many the first quarter.*

*'**No one will have a computer at home**.' My wife said when I tried to buy one for home in the mid '80s.*

However, Ken Olsen, founder of Digital Equipment Corporation, in 1977, said it first. **"There is no reason anyone would want a computer in their home."**

Darryl Zanuck, executive at 20th Century Fox, 1946: **"Television won't be able to hold on to any market it captures after the first six months. People will soon get tired of staring at a plywood box every night."**

Alex Lewyt, president of Lewyt vacuum company, 1955 **"Nuclear-powered vacuum cleaners will probably be a reality within ten years."**

Richard Feynman a physicist on the Manhattan Project that built the Atomic Bomb wrote in his biography that at the end of WWII, lawyers came around and pestered all the scientists to write down possible uses of nuclear power so they could file patents. Feynman tried to avoid the task, but eventually wrote down as many absurd things as he could as an act of defiance. His name is on the patent for the nuclear-powered airplane. Strangely enough, they tried to build one before realizing all the horrendous problems that Feynman intuitively understood. But to my knowledge, Feynman did not try to patent the nuclear vacuum cleaner.

Robert Metcalfe, founder of 3Com, 1995 **"Almost all of the many predictions now being made about 1996, hinge on the Internet's continuing exponential growth.**

But I predict the Internet will soon go spectacularly supernova and in 1996, catastrophically collapse."

3Com was an early leader in the computer networking business. By 1997 they merged with US Robotics that made computer modems and then in 2010 was absorbed by HP. Perhaps their lack of belief in the growth of the Internet led to their downfall?

Nathan Myhrvold, former Microsoft CTO, 1997 **"Apple is already dead."**

Bill Gates, founder of Microsoft, 2004 **"Two years from now, spam will be solved."**

My favorite quote from Bill was his pronouncement that personal computers "**would not need more than 640,000 bytes of memory**". The engineers at Intel, knew that wasn't true when we got IBM to use the Intel chip. We told Bill and his team not to make ten basic assumptions on a list we provided in designing the DOS Operating System. To my knowledge, Bill and team broke eight of the ten.

Bill spends time thinking about the future. Which goes to show how hard it is to predict. In 1989, he said "**We will never make a 32-bit Operating System**" (Windows) Not only did Windows quickly go to 32 bits, it is 64-bits today.

Sir Alan Sugar, founder of the electronics firm Amstrad and member of the House of Lords (UK), in 2005, **"By next Christmas, the iPod will be dead, finished, gone, kaput."**

Figure 56 iPod Wikipedia.org

Now this was either stupid or on the wrong time scale. Certainly, sales of iPods were strong in 2006, but ten years later, dedicated iPods are not as popular. Music has moved on to our smart phones and to a streaming model. If you are going to make a prediction, you do well to leave off specific dates.

I did not repeat some of Bill Gates other comments. As CEO and partner with IBM he likely had to publicly support products that he had personal qualms about. And Bill is a very smart man. As I said predicting the future is hard.

You can however look at trends. Gordon Moore was a founder of Intel. In 1965, he predicted that the number of devices on an integrated circuit would double every 18 months "for the next 10 years". It's been over 40 years and this prediction is still hanging on. Its death has been predicted several times. Other trends are Disk drive capacities double every year and Fiber Optic capacity doubles every nine months.

When I was at Intel, we convinced ourselves that computer hard disk drives that had spinning disks like a record player, would soon (1981) hit a wall and no longer increase in memory size. They were then about 5 to 20 Mbytes in the 19-inch size. Intel came up with Bubble Memory. We were ready to sweep in and take over the market. Trouble is, no one told the hard disk guys they were dead and they kept on doubling memory size every year right on schedule for the last several decades. A 2-inch hard disk is now about 2 Terabytes (1,000 Gigabytes). Bubble memory turned out to be unreliable and died a quick death. Its demise made me more skeptical about future predictions. However, if you wait long enough the general idea often comes true. The computer I'm using to write this book has FLASH memory and no hard drive. So, solid state memory is replacing the spinning disk. Just several decades later, and not bubble memory.

What does that all mean? Back in the mid 90's, Wired Magazine was predicting that the price of data on the Internet fiber-optic backbone was falling rapidly and would soon be free. I'm sure the author put some caveats on his prediction but the basic premise was network bandwidth is getting cheaper. And it still is today. What has changed since the mid 90's is that in the US and Canada, you don't care where you are calling. Calling California from Georgia is not more expensive than calling next door. It means you can stream music, download software and watch Netflix's "House of Cards" in UHD, streamed over your internet connection. Calls to family in Germany are two cents a minute using a Voice over IP service. If we both use VoIP, it is free

---

We learned at Intel that voice recognition in industry could succeed or fail based on how it was perceived by the workers using it. If it made the workers job safer and easier, it was a success. If it allowed for 30% reduction in the workforce, the workforce made it fail. Sometimes it's not about the technology.

(assuming we both have high speed internet connections).

When my company was building Digital Video Recorders, we paid close attention to the exponential increase in technology. Moore's Law caused me to file a patent on voice recognition in set top boxes. Today Apple, Google, and Roku among others have this as a fundamental part of their current offerings. All those extra transistors allow you to do science fiction type stuff.

The problem with the future,

is that it keeps turning into the present.

Bill Waterson of Calvin and Hobbs

---

*OK, enough excuses. Time to put my money
where my mouth is. My predictions.*

- *LED light bulbs rule.*
- *Most people will connect their devices to
  the Internet using Wi-Fi.*
- *Wi-Fi will be commonly available in any
  reasonably populated area.*
- *Average Internet (network) speeds will
  increase from 20 Mbps today to close to
  100 Mbps. 1,000 Mbps will be common in
  large metropolitan cities.*
- *Your next TV will be UHD (4k). You
  cannot perceive the extra pixels but the
  color will be great.*
- *8K TV will never take off for home use.*
- *Sound in the home will not exceed 7.2.4.
  Theaters might use 22 channel sound.*
- *More and more movies and video shows
  will use High Dynamic Range Color (HDR).*
- *More grandparents will read to their
  grandkids over Skype, Apple Face Time or
  Google+*
- *More music will be consumed over
  streaming media. But it may not be
  consumed by you or me. I like my iPod
  and owning music.*
- *AT&T and Verizon will lose more and more
  wireline customers as people move to
  mobile and Voice over IP. In time, they
  will have a completely different business
  model or they will be extinct.*

- *More people will give up on dedicated cameras as smart phone cameras get better and better.*
- *The trends show more people getting news from Facebook and that is just fundamentally wrong. (Sorry)*
- *More people will be hacked through IoT before IoT manufacturers figure out how to have good security.*
- *Someone will find a problem for Internet of Things (IoT) to solve. Maybe it is not one thing but a bunch of things.*
- *Moore's Law will keep chugging for a little while longer.*
- *Spinning Hard Disk drives will be replaced by solid state memory in more and more applications.*
- *The price of sending data around the world will keep dropping. By the way, voice, music, video is all just data.*
- *Your cable TV system is likely* all digital *today. It will move from Broadcast to On-Demand as it moves to all IP. (Internet Protocol or packets of data) Sports will still be broadcast.*
- *Video will be offered in a wider variety. Consumers will opt for packages with fewer channels. There will be more choice. The competition will increase and the cost will drop. These packages may or may not come from today's service providers*

And that about wraps it up

Notes: